The
Practical
Vibration
Primer

The
Practical
Vibration
Primer

Charles Jackson

Gulf Publishing Company
Book Division
Houston, London, Paris, Tokyo

Acknowledgments

To Jean, for her
love and understanding.

The autnor would like to acknowledge Monsanto's permission to write this book and its assistance in providing data and illustrations.

Further, engineers and technicians in the Mechanical Technology Section who contributed information include:

Howard Blackburn, Senior Engineering Specialist
Carl H. Duhon, Senior Engineering Specialist
James H. Ingram, Senior Engineering Specialist
James L. Glessman, Senior Engineer
Charles R. Brown, Engineering Specialist
James L. Smith, Technician
Tony P. Spruiell, Maintenance Superintendent (formerly of Mechanical Technology Section)
Marvin W. Ringer, Jr., St. Louis

In the technology exchange concept, while I have tried to exchange my experiences freely, I always derive more from others in the same or related fields. To name a few is wrong for fear of slighting others; however, I must thank the following for directly or indirectly contributing to this book:

John S. Sohre, Consultant
John S. Mitchell, Consultant
Donald E. Bently, President, Bently Nevada Corporation
Robert C. Eisenmann, Consultant, Bently Nevada Corporation
Roger Harker, General Manager, Bently Nevada Corporation
Glen Thomas, President, IRD Mechanalysis
"Pete" Bernhardt, IRD Mechanalysis
John R. Harrell, IRD Mechanalysis, Baton Rouge
Terrell Jamison, formerly of Tekronix
A.J. Campbell, Dresser Industries
Dr. Edgar J. Gunter, Jr., Consultant
A. Steward Maxwell, Ontario Hydro, Canada
Vern Maddox, General Electric
Jon Palm, Dymac
Peter Sundt, President, Metrix Corporation
John W. Woodworth, Bently Nevada Corporation, Germany

The Practical Vibration Primer

Library of Congress Cataloging in Publication Data

Jackson, Charles, 1929-
    The practical vibration primer.
    Includes bibliographies and index.
    1. Vibration. I. Title.
TA355.J28   620.3   79-50249
ISBN 0-87201-891-1

First Printing, May 1979
Second Printing, September 1981

Preface

This book is called a primer because that is what it is. The idea of the title was derived from *The Strain Gauge Primer* by Perry & Lissner. Their primer was valuable to me when I was learning application of strain gauge (art, science, and techniques) about fifteen years ago. Almost everything else that I read was too complicated. When I could understand the material I read, I found it offered no immediately useful information.

The inspiration to write the book was derived from many calls each year from people trying to learn rotating equipment and its behavior, specific vibration analysis, or both. The complaint was the same, "Where can I read something that I can understand and get help from immediately? Everything I read is loaded with differential equations, which are of little help."

It is hoped that this primer can also be used as a supplementary book to universities in graduate and undergraduate courses, particularly those offering courses in turbomachinery, vibration, mechanical machinery lab courses, rotor dynamics, industry assistance analysis contracts, or instrument related courses in EE or ME.

Many people are not properly introduced to a machine in the field. Portions of this book may ease that apprehension. The first eight chapters were released consecutively by the publisher in magazine form. Chapters 9-12 cover areas of related interest. Alignment contributed to the solutions of over half the vibration problems during the first five years of analysis work. Chapter 10, was written for maintenance people having difficulty setting thrust bearings on turbines. The calibration of rotor displacement sensors was added as it is difficult to understand; yet a thrust failure is very severe and costly. Orbit balancing, Chapter 11, was derived from an ASME article written in 1970; it has been heavily quoted by universities doing lab studies. Chapter 12 touches on the need to have a group dedicated to the analysis, protection, redesign, and selection of rotating machinery. An annual 10 percent improvement in the maintenance expenditures is expected within the first five years of a medium to large operating company. These companies may have a maintenance expenditure of $10 to $15 million per year. Rotating equipment will generally fall about 25 percent of that total expenditure. For a starter, $250,000 is not bad. This commitment requires a good balance of real estate, equipment, and personnel.

If there is a bottom line to troubleshooting from the experience of the author, it is this— "Most problems resolve to be an infringement against the basic fundamentals of good design or poor maintenance practices without good basic fundamentals."

C. Jackson
March 1979

Contents

Basic Motion-Displacement, Velocity, and Acceleration

Vibration exists when a system responds to some excitation, generally referred to as a forcing function. If the weight in Figs. 1-1A and 1-1B is displaced by a force and released, the motion will either die or continue. If the motion quickly decays, no harmful continuous vibration would ever exist. If motion continues, vibration continues, and the severity must be appraised because many harmless vibrations exist. Consider the fluctuations of an airplane wing in flight or the smooth action of an automobile's springs while traveling down a road.

Vibration parameters can be expressed as force, displacement, energy velocity or strain just to name a few common units. Rotating machinery vibration will usually be expressed in terms of displacement, velocity or "g's" of acceleration. These units all refer to the *amount* of vibration. Vibration frequency is many times more important. Phase of vibration motion is also seldom considered, but can be the most important factor in solving a problem.

Pure harmonic motion is probably the best starting point in understanding vibration. This motion is synchronous with, or an even multiple of, a synchronous frequency. Synchronous motion implies that the vibration frequency of a structure or machine is equal to the primary forcing functions, such as the rotating speed. Motions may be axial, radial or torsional. Radial motion

will now be discussed to develop an understanding of the terminology.

Motion of the weight in Fig. 1-1B is viewed as an up and down motion but it can also be expressed in circular terminology (Fig. 1-1C). As the weight moves from point 0 to 1 to 2 to 3 to 4, it can be noted that 0, 2, and 4, are the same points of vertical motion. If the rotating shaft is viewed as the same vertical motion, points 0, 2, and 4 would express the same relative positions. If the cycle from 0 to 4 takes place in one second, the frequency is 1 cycle per second, or simply 1 cps or 1 Hertz (named after the German physicist, Heinrich Rudolph Hertz). The *period* of a motion is the time required for a point to return to the same repeatable position in motion; in this case, it is 1 second. Period is the reciprocal of frequency, in this case 1 sec./cycle. Frequency of electrical current or voltage in the United States is 60 Hz, 60 cps or 1/60 sec./cycle. Peak motion in terms of displacement from rest is obviously from 0 to 1 or 2 to 3 or 3 to 4. Because symmetrical form of motion is under discussion, it does not matter which is used providing the motion is steady and not increasing or decreasing. If peak terms are used in vibration, the motion from zero to peak is implied, 0 to 1 in this example. For peak-to-peak motion, the movement from 1 to 3 is implied. The velocity, or change in

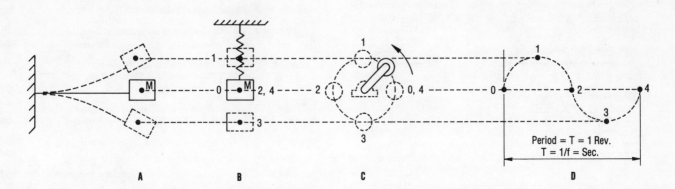

Fig. 1-1—Types of vibratory motion.

distance per unit of time (change of x per unit of t, i.e., dx/dt), occurs not at point 1 because the motion comes to a halt here and actually reverses, but at point 2 or 4.

If the above described motion was expressed as circular motion, the mass of the rotor, which may be displaced some distance r from the center of the shaft, whirls in a circular path, or orbit, as each revolution is completed. The cycle now becomes a revolution which is completed in a unit of time. If the shaft rotates 60 revolutions per 1 second, then the frequency would be 60 revs./sec., 3,600 revs./min. (rpm). The time per cycle or revolution would be 1/60 or 0.0166 sec./cycle. The time of one cycle or revolution can also be expressed in degrees, i.e., 360 degrees per revolution, or radians (2π radians per revolution). Radians are dimensionless mathematical expressions used in vibration calculations. Angular velocity, ω, is often expressed in radians/sec. Circular frequency, in cycles per second, can be expressed in radians/sec. by multiplying by 2π. Circular frequency is synonymous with circular or angular velocity, i.e., $\omega = 2\pi f$. Converting from machine speeds, stated in rpm, to radians/sec. is accomplished by multiplying (rpm) $(2\pi/60)$. You can estimate this by simply moving the decimal place one position to the left, e.g., 3,600 rpm is approximately 360 radians/sec. (exact = 376 radians/sec.).

Now, you can express a certain phase of motion if the frequency and the time are given. For example, keeping the frequency in mind and starting at point 4, where the mass of the rotor has traveled in 2 milliseconds (0.002 seconds)? Multiplying the frequency by the time yields (60 cycles/sec.) (0.002 sec.) = 0.12 cycles or approximately 1/8 of one cycle or in radians, $\omega t = (2\pi 60)$ (0.002 sec.) = 0.75 radians (43.2 deg.).

Referring to Fig. 1-2, the displacement at any period of time from the described neutral axis, is represented by x. Recalling from trigonometry the sine of the angle $\omega t = x/A$; where A represents the amplitude (zero to peak value) and ωt represents the angular position at time t. If speed of rotation was 3,600 rpm, 60 rps, 376 rad./sec. and the time was 3 milliseconds, then ωt would be (376 radians/sec.) (3) (10^{-3} sec) = 1.128 radians or 64.63°. If the amplitude equals 1 mil (0.001 inches) the displace-

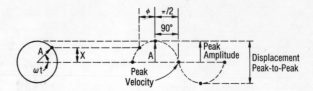

Fig. 1-2—Relationship of vibration amplitude and angular position.

ment, $x = A \sin \omega t = 1 \sin 64.63° = 1$ (0.903) or about 0.9 mils. This is often referred to as an instantaneous value of displacement, but the finite value of ωt is needed. When you refer to the peak value, ($\sin \omega t$) is assumed at maximum value (1) and $x = A(1) = A$ or 1 mil O/P (zero to peak).

Later, it will become apparent that displacement, velocity and acceleration will be thought of as quantitative terms. Frequency will have a strong bearing on the source of vibration (particularly so in diagnostic work). And phase will be helpful in being more specific as to where a trouble lies within a revolution or in balancing, where the high spot or heavy spot is located relative to the vibration sensor.

A simple exercise in estimating critical speed will be used to illustrate these principles. In Fig. 1-3A, a 1,000-pound disc is supported on a 2-inch diameter steel shaft (massless) rigidly supported by bearings on 36-inch centers. Neglecting the bearing supports for the moment, i.e., assumed ultimately rigid, the critical speed, ω in rad./sec. can be determined from the expression in Fig. 1-3B, $\omega = \sqrt{K_s/M}$.

If the 1,000 pounds are statically placed on the 2-inch diameter shaft, the deflection at the center will be:

$$Y = \frac{W L^3}{48 E I}$$

W = weight, 1,000 pounds (shaft is assumed massless)
L = bearing span, 36 inches
E = modulus of elasticity, 30,000,000 psi for steel
I = moment of intertia of the shaft, in.$^4 = \dfrac{\pi d^4}{64} = \dfrac{\pi (2)^4}{64}$
$= 0.785$ inch^4

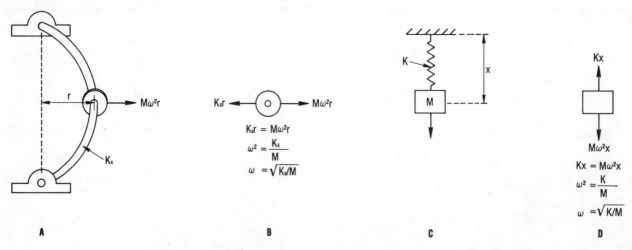

Fig. 1-3—Forces and resonant frequency of vibrating systems.

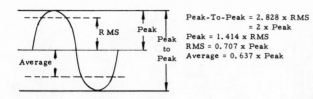

Peak-To-Peak = 2.828 x RMS
= 2 x Peak
Peak = 1.414 x RMS
RMS = 0.707 x Peak
Average = 0.637 x Peak

$g = 386.089$ inches/sec^2 = 9.80665 M/sec^2
Sound Power = $W_o = 10^{-12}$ Watts
Sound Pressure = $P_o = 20 \mu Pa = 0.0002 \mu$ bar
1 bar = 14.5 psi = 10^5 Pa (N/M^2)
mil = 0.001 inches = 25.4 μ M
Velocity = $V_o = 10^{-8}$ M/S
dB (Power) = 10 Log (W/W$_o$)
dB (Pressure, Voltage) = 20 Log (P/P$_o$)
= 20 Log (E/E$_o$)

Motion	Symbol	Units									
Acceleration	A, $\ddot{x}$	in/sec², RMS	386.09	272.96	136.482	628.32	444.22	222.11	394.58	278.97	139.55
	A, $\ddot{x}$	in/sec², Peak	545.93	386.09	193.045	888.44	628.32	314.16	557.58	394.58	197.392
	A, $\ddot{x}$	in/sec², P/P	1091.86	772.18	386.09	1776.88	1256.64	628.32	1115.88	789.168	394.78
Acceleration	A, $\ddot{x}$	g, RMS	1.0	.707	.353	1.627	1.15	.575	1.022	.772	.361
	A, $\ddot{x}$	g, Peak	1.414	1.0	.5	2.3	1.627	.814	1.445	1.022	.511
	A, $\ddot{x}$	g, P/P	2.828	2.0	1.0	4.6	3.254	1.627	2.89	2.044	1.022
Velocity	V, $\dot{x}$	in/sec, RMS	.614	.434	.217	1.0	.707	.3535	.628	.444	.222
	V, $\dot{x}$	in/sec, Peak	.869	.614	.307	1.414	1.0	.5	.888	.628	.314
	V, $\dot{x}$	in/sec, P/P	1.738	1.228	.614	2.828	2.0	1.0	1.776	1.256	.628
Distance (Magnitude Or Displacement)	D, x	mils, RMS	.978	.691	.346	1.591	1.125	.563	1.0	.707	.353
	D, x	mils, Peak	1.383	.978	.488	2.25	1.591	.796	1.414	1.0	.5
	D, 2x	mils, P/P	2.766	1.956	.978	4.5	3.182	1.591	2,828	2.0	1.0

Note: All values above were determined at 100 hz (6000 cpm).

Fig. 1-4—Relationship of displacement, velocity and acceleration units.

the deflection is then:

$$Y = \frac{(1,000 \text{ lbs.}) (36 \text{ in.})^3}{48 (30,000,000) \text{ psi} (0.785) \text{ in.}^4} = 0.0413 \text{ in.}$$

the spring constant of the shaft is:

$$K_s = \frac{1000 \text{ lbs.}}{0.0413 \text{ in.}} = 24,228 \text{ lbs./in.}$$

The spring constant is explained this way for a better feel for the flexibility.

An easier solution for the spring constant is:

$$K_s = \frac{48 \text{ EI}}{L^3} = 24,228 \text{ lbs./in.}$$

Mass of the disc equals the weight divided by gravity, or W/g, where $g = 386$ in./sec.^{2}, so:

$$\text{mass} = \frac{1,000 \text{ lbs.}}{386 \text{ in./sec.}^2} = \frac{= 2.59 \text{ lbs.-sec.}^2}{\text{in.}}$$

The critical speed is then:

$$\omega = \sqrt{\frac{K_s}{M}} = \sqrt{\frac{24,228 \text{ lbs./in.}}{\frac{2.59 \text{ lbs.-sec.}^2}{\text{in.}}}} = 96.70 \frac{\text{radians}}{\text{sec.}}$$

or $(96.70 (60/2\pi) = 923.5$ rpm

Note the similarity between this situation and Figs. 1-3C and 1-3D where a spring supports a mass and is put into oscillation. If the spring constant is 24,228 lbs./in. and the weight is 1,000 lbs., the same critical frequency or resonance occurs. As a matter of fact, this is the preferred method of measuring springs for their constant. Put them in resonance, count the cycles per second, and solve for the spring constant, K_s.

A final chart (Fig. 1-4) is provided for reference and study. A frequency commonly used to calibrate vibration analyzers and sensors is 100 Hz (cps). Because this corresponds to 6,000 revs./min. for rotating machinery, it seems a reasonable calibration point. Another point near the intended measurement frequency is also preferred, plus a calibration curve showing the deviations in signal at all frequencies within the guaranteed range.

This chart displays acceleration, velocity and displacement in the three most commonly stated units. It is well to study these units. If they are not understood, a measurement can be significantly in error. For example, if peak-to-peak and rms values are confused. Equivalent values can be read vertically with each column holding one value, e.g., mils peak-to-peak at unity (1).

Sensors

Sensors of some type must be used to measure vibration. Proper choice is important though there is some overlapping in frequency response. Frequency response simply means the range of frequencies over which the response of the sensor can be considered linear or at least predictable. Many instrument systems are available to condition the signal taken from the vibration sensor. These systems will not be discussed at this time. Here is what you should know about the sensors.

In my opinion, there is no one sensor for all jobs. This point will be proven in the discussion of sensors. Reference is made to certain manufacturers' data on sensors. Many different sensors of the types described are used. There are other reputable sensor manufacturers, but I do not have personal experience with their line.

The more practical vibration sensors used with machinery and structures will be discussed, i.e., distance or gap measure, velocity and acceleration. This is a *practical primer* so only the more conventional will be discussed; and that decision involves environment, range, size, adaptability, cost, acceptance, accuracy, metallurgy and many other considerations.

DISPLACEMENT SENSORS

While there are linear differential transformers, magnetic, capacitance, reluctance, proximity sensors and hosts of others, the sensor most used is the eddy current proximity sensor. This sensor generates a path or eddy current field to *any conducting* surface and the available energy in the form of output voltage is drained as the gap between sensor and conducting body is reduced. Input voltage, by a recent Vibration Monitor Specification, API 670, by the American Petroleum Institute, will control this input DC voltage to 24 vDC. Calibration curves will show an output of 100 millivolts (peak-to-peak) per mil (0.001 inch) peak-to-peak. The metric equivalent is 4 mv/micro-meter. The other preferred calibration is 200 mv/mil or 8 mv/pM. *The displacement proximity sensor measures gap and nothing else.* It is well to understand this simple fact. Further, in phase work, such as balancing, it defines only *high* spots (low gap or simply the closest the shaft is to the sensor), never *heavy* spots.

Advantages:

● Small in size, normally 190 to 300 mil tips on ¼-inch-28 or ⅜-inch-24 threads

● Inexpensive, on the order of $50, for sensor only; over $200 with oscillator-demodulator

● Unaffected by lubricating oil and most gasses

● Can sense shaft motion relative to bearing if mounted to the bearing

● Temperatures of 250°F give satisfactory service. Higher temperatures cause DC voltage shift and loss of linearity at the end points in range.

● Frequency response of DC (zero frequency) to 5 kHz (300,000 cycles per minute)

● Shaft attitude in the bearing as well as shaft orbits can be plotted

● Same sensor can measure rotor thrust (axial) displacement. Scope display (AC coupled) of this signal for axial vibration and thrust collar runout is often helpful, e.g., a sudden change in thrust collar runout could indicate a thermal shaft bow on a turbine start-up or change of steam conditioning, e.g., slug of liquid

● Sensor can be used for phasing marks or shaft positioning, e.g., cutter blades or machine tool limitors

● Sensor can be seismically supported for absolute and relative measure

● The oscillator-demodulator output is good for over 1,000 feet of transmission

● Can be used for impedance or resonance test, e.g., compressor cover resonance (shaker, impeller, series of probes on support bar), where the effect from sensor attachment might alter the results.

Disadvantages:

● Sensor is susceptible to changes in shaft surfaces, e.g., mechanical runout, electrical runout, coatings of different conductivity (chrome plating), shaft finish.

• Submergence in water can be troublesome unless well sealed in ceramic even though water is basically inert.

• There are limits in temperature and frequency response

• The DC bias (probe gap) voltage must be blocked or "bucked" if an FM recorder is used on the raw vibration signal

• Is not self-generating, i.e., requires a separate source of power to operate

• Is susceptible to induced voltage from other conductors, e.g., 110 v, 60 Hz placed improperly in the same conduit with the proximity sensor signal cable.

Pitfalls. There are several pitfalls with any device. Most pitfalls with proximity sensors have to do with proper instrumentation installation and care of cables. A mechanical pitfall is a probe holder or body that is a resonant structure. One should attempt to provide holders that are ⅜-inch minimum cross section. Result of a weak, e.g., ⅛-inch angle, support is that you may detect a ghost frequency from a resonant probe holder that does not relate to the machine under analysis. My first experience with the problem was a machine that showed a 24,000 cpm (rpm) frequency component down the entire train. Solution was to raise the holder resonant frequency by using heavier supports. Reading the outboard bearing shaft motion on a machine by reaching over the thrust bearing with a support section, often 1½ inches offset, is an area consistently overlooked by many machine builders. Gussetting the existing brace often-times corrects the problems with minimum delay.

Holding the proximity probes with tangential clamps is much preferred to single or double jam nuts. An eager craftsman is apt to multiply one good probe into two bad ones with one or two wrenches.

Failure to provide proper grounds at instruments *and* sensors can result in 60 Hz and 120 Hz ground loops.

VELOCITY SENSORS
(Also called geophones and seismic sensors)

This type of sensor transmits a voltage proportional to velocity from the movement of an armature (coil of wire on core) seismically mounted in a magnet. Thus, the magnet moves in and out about a coil that remains in one position. One might think of this as a reciprocating generator. This type of sensor is generally placed at a point where the force is expected to be great, e.g., at the machine's bearings. If you were to place this sensor at a less sensitive location, e.g., the b a s e p l a t e ; baseplate, ground or piping vibrations may appear as the predominate signal. The signal is quite strong from this sensor and can be transmitted some distance (over 1,000 feet in some cases) without amplification.

The sensor can be supported with a magnet, bolted to the bearing, held with epoxy or dental cement, just to name a few methods. There can be several false resonant signals, e.g., not of the machine but of the mounting, if you are not careful of proper mounting details. For continuous monitoring, I have felt content using a spot-faced machined area coupled with a fine (NF) machine thread and a suitable epoxy cement. On short measurements, I never use flimsy temporary attachments and rarely "hand hold" the sensor. I prefer to use a good 14 mil magnet for temporary sensor attachment (implies that vibration over 14 mils exceeds the holding power of the magnet and often the analyst).

Some velocity sensors weigh about 600 grams (typical of the IRD 544). This particular model mounts on a ¼-inch 28 tapped stud, and has a resonant frequency of 16 cps *critically damped* at 77°F (that means it is not supposed to amplify at 16 cps or 960 rpm). Sensitivity is about 1.08 volts (peak) per inch/second (peak) in velocity units. From Part 1 of this series, one would expect a display on an oscilloscope of 2.16 volts peak-to-peak. Further, if the calibrated scope gain knob was set on 1 volt per cm (division) then a 2.16 cm peak-to-peak signal would be observed.

Carrying this a step further, if the IRD 544 velocity pick-up was being shaken at 1 inch/second peak velocity at 100 cps (6,000 cpm) the frequency would also be discernable on the scope. Setting the time base of the scope on 10 milliseconds/cm, a positive peak of the sine wave would appear every centimeter on the horizontal sweep (Fig. 1-2).

$$\frac{100 \text{ cycles}}{\text{sec.}} \times \frac{10 \ (10^{-3}) \text{ sec.}}{1 \text{ cm}} = \frac{1 \text{ cycle}}{1 \text{ cm}}$$

The displacement (peak-to-peak) of the above situation would be 3.18 mils (0.00318 inch). You may recall that to obtain the amplitude o/p (zero-to-peak) from the velocity simply divide by the frequency in radians/second or $2\pi f$:

$$\frac{(1 \text{ in./sec. x } 1{,}000 \text{ mils/in.})}{2\pi 100 \text{ rad/sec.}}$$

= 1.59 mils peak (o/p) of 3.18 mils p/p.

Looking at Fig. 1-4, this could have been determined (since 100 Hz is used) by reading down from 1 inch/second peak in Column 5 to read 3.182 mils p/p.

Advantages:

• Strong signal

• Self-generated voltage

• Stable below 30 g's, less than 500°F

• Rugged construction

• Can mount to shaft or read shaft with fishtail

• Mounts in any position (generally), i.e., omnidirectional

• Transverse response is less than 5 percent to 1kHz (this means the response at right angles to the direction it is pointing)

• Reasonable accuracy to 300,000 rpm

• Adaptable to direct probe.

Disadvantages:

• Heavy and large in size, i.e., some mounting would fail the part to be measured

• Ambient accuracy at ± 8 percent limited to 1,000 cps

• Output signal decays exponentially below 10 cps but can be corrected.

• Price over $500.

ACCELEROMETERS
(Acceleration sensors)

I will limit my discussion to the piezoelectric accelerometer which supports a mass on a piezoelectric crystal (quartz or others). This type crystal produces a charge, or voltage, if the crystal has a force applied to it. Just to illustrate outputs, an Endevco 2233E has a charge sensitivity of 60 pico-coulombs/g (nom.); voltage sensitivity is 45 mv/g nominal (with 300 pico-farads of external capacitance); transverse sensitivity is 3 percent max. (down to 1 percent on specified orders); sensor's capacitance is 1,000 pF (± 20 percent); weighs 32 grams and mounts on a 10-32 stud, and mounted resonance is 32 kHz and the frequency response is 2,000 to 6,000 Hz (± 5 percent). Endevco will give a National Bureau of Standards traceable curve to 5,000 Hz and can calibrate to 10,000 Hz. (Note: A guide to *good usable* frequency range would be to take 1/5 of the resonant frequency.)

At this point, a considerable amount of new numbers has been introduced. The accelerometer is more complicated to understand but necessary if high frequencies are to be measured or lightweight sensors are needed. These units are also quite useful at low frequencies.

There is a formula that you need to remember in using accelerometers, i.e., Q(pico-coulombs) = C(pico-farads) V(volts). I will put this in terms of importance to determine voltage output for 1 g rms of shake (1 g = 386.089 inch/second^2 or 9.80665 M/second^2). The above voltage output, 45 mv/g, is conditional on 300 pico-farads (300 x 10^{-12} farads). (Note: The term pico has replaced micro-micro and either way, that isn't just a whole bunch of Farads.)

If the charge sensitivity is divided by the total capacitance, then the voltage sensitivity should check out ($V = Q/C$). Total capacitance is the 1,000 pF of the sensor plus the 300 pF of external capacitance or 1,300 pF total. Dividing this into 60 pC yields 46.15 mv/g so there must be an extra 33 pF of cable or connectors involved in order to get 45 mv/g (exact).

If the accelerometer does not have a charge convertor or charge amplifier very close to the 10 to 11 feet of cable (cable at 30 pF/ft.), then the output will drop off quickly. Thus, one must add the price of convertors to the price of accelerometers for data taken, plus some form of power supply.

Advantages:

• Wide frequency response

• Lightweight

• Good temperature resistance; 1,000°F on special order

• Moderate pricing.

Disadvantages:

• Sensitive to mounting and torque

• Sensitive to the degree that you can get too much data

• Resonance can be excited in the sensor often requiring a low pass filter. (This is a filter that passes low frequencies but not high frequencies.)

CONCLUSIONS

• Most sensors, whether velocity, proximity or accelerometer, will cost about the same for the necessary equipment to send a signal. This will range from $300 to $500 each, depending on the particular style and manufacturer.

• Velocity sensors are bulky, hard to fit in small spaces, self-generating with strong signals, but limited frequency range attenuates signal on the low end frequencies. Their use is questionable on high speed machinery with over 10:1 total weight/rotor weight ratios, definite above 20:1, if the bearing housing is not separate, e.g., a barrel type centrifugal compressor. Velocity is the best single medium frequency parameter to measure, but it is not necessarily the best type of sensor.

• Accelerometers can easily be 2 grams to 45 grams. They can be used for small spaces, covering low and high frequency range and high temperatures. They are sensitive and may provide too much noise. Impedance matching is needed; accelerometers would be effective on gearing, blade pass frequencies studies, shaker calibration, impedance studies and systems which are light in construction, such as a thin membrane, e.g., a turbine exhaust header.

• Proximity sensors have many uses. They suffer from shaft surface effects; however, the ability of early detection of shaft excursions within the bearing clearance is very important. Studies in shaft orbits are often referred to as eccentricity ratios. This is the ratio of radial motion of the shaft or journal relative to radial clearance of the bearing, e.g., an eccentricity ratio of 1 would define a shaft hitting the babbitt of a journal bearing—the total radial clearance would have been removed by the shaft motion.

• Proximity sensors are particularly useful for tests where the sensor should not contact the element being measured. They are effective in determining attitude of a shaft within a bearing for possible location of anti-oil whirl stabilization pressure dams; or motion of a main shaft relative to sealing elements, particularly in studies of possible seal rubs.

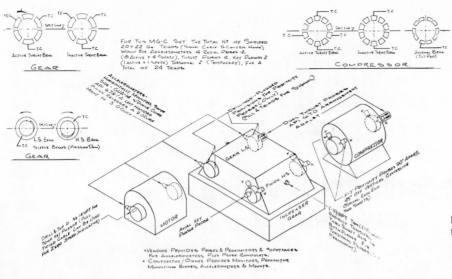

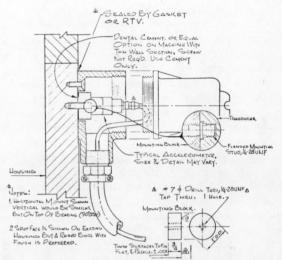

Fig. 2-1—The API 670 Non-Contacting Vibration and Axial Positioning Monitoring System.

Fig. 2-2—The API 678 Accelerometer-Based Vibration Monitoring System.

I am often asked, "What kind of analyzer or sensors do I need at my plant (or in my consulting business)?" My traditional answers continue to be "What are you going to measure? What are you looking for? How is the machine built? What must be your display? How important is time to you?"

I should conclude by covering an obvious omission from this particular section. *The more you measure, the more you question the data.* Most people question the data taken by others. Do not accept data "carte blanche" if not taken by you, unless you have a degree of confidence in the people providing the data. Please provide some means of checking calibration on the sensor and sensor's connecting system to the read-out device. Our department uses two distance (gap) calibrators: one fixed frequency battery powered shaker at 1 g to 300 gms and 100 Hz, and one variable frequency, variable force, variable position shaker with an accompanying power amplifier which incorporates a reference accelerometer on the shaker table.

API 670, NonContacting Vibration And Axial Positioning Monitoring System, was released by the API June 1976, and has been accepted worldwide as a specification for installing monitors for displacement (eddy current) sensors. Also, it was the first such document written by the subcommittee of CRE (Committee of Refinery Equipment) to go further than a purchase specification. It projects into the area of installation, including electrical requirements, wiring, numbering, orientation, mounting, and protective conduit/condulet systems. See Figure 2-1.

API 678, Accelerometer - Based Vibration Monitoring Systems, dated May 1981. This specification was inspired by the success of API 670, but for seismic accelerometers. The piezoelectric accelerometer is chosen as the basic sensor, yet allowing for the signal to be integrated to display velocity. See Figure 2-2.

The author installed seismic sensors per API 678 on the motor and low-speed gear bearings of two motor—gear driven centrifugal compressors which were commissioned successfully in October 1980. Several gears, screw compressors, fans, and pumps have been accommodated in the year prior to the second printings of this book. An outline for two motor—gear compressors is shown here to illustrate the joint intallation of seismic and displacement (radial and axial) sensors.

Logarithmic Scaling
and Filters

There are two types of scaling used in vibration work. One is linear and the other is logarithmic. The linear is quite easy to use and is straightforward. A measured value can be compared with others without much study because values, for example of one-half, will be one-half value or one-half the peak on a recording. However, when you switch to log scaling this easily discernible linearity disappears. On the other hand, log scaling can put a lot of information of varying values into a small space and often enhances the presentation from a signal-to-noise ratio standpoint. Log scaling along with averaging can improve the signal-to-noise ratio. Having a good signal-to-noise ratio implies that low signals are detectable over the base noise of an instrumentation system. Here the noise should be referred to as electrical noise and not acoustical noise.

Most people that I have talked with have an immediate fear of log, decibel (dB) terminology. While I find it is more difficult to translate quickly, it is not that bad a system. I am of the opinion that people look at decibel (dB) values as something mysterious. The first reaction is to think of sound measure. Though the acousticians use dB scaling, the term first originated in physics and primarily in electrical terminology.

Technically, decibels are, in addition to being $\frac{1}{10}$ of a bel, a logarithm equal to 10 times the log to the base 10 of the ratio of two power levels.

dB = 10 log (power level/power reference)

or

dB = 10 log (power output/power input)

The upper expression is used in power distribution systems and if a supply system has a 10 dB gain then this amounts to a 10:1 gain. If the reference power was 100 kw, then the output is 1,000 kw:

dB = 10 log (1,000/100) = 10 log (10) = 10 (1) = 10.

If an amplifier had a gain of 20 dB, then this amplifier would output 10^2 or 100 times its input.

If the sound power level base is 10^{-12} watts, then 100 dB would represent a PWL of 0.01 watts. Said another way, 100 dB in power reference implies a power ratio of $10^{10}:1$ (ref.).

Now you may be asking yourself why sound pressure levels, voltage levels and current levels are referred to on a 20 log basis, i.e., dB = 20 log (volts measured/volts ref.). Power varies as the square of voltage with the impedance (resistance) the same or constant. Power (watts) = E^2/R, then dB = 10 log (E measured/E ref.)^2, but logarithms of numbers raised to an exponent equal the exponent times the log; so, dB = (10) (2) log (E measured/E ref.) = 20 log (voltage ratio).

When you hear two people discussing a gain of 40 dB, one may be thinking 100:1 and the other 10,000:1. The first (100:1) is on a voltage basis, the second (10,000:1) is on a power basis.

Has this exercise been meaningful in a primer on practical vibration? I think so, but you may be in doubt. First, dB or log scaling is simply that, a scale and nothing more. It does not mean anything until you define the reference. If a recorder has a 40 dB potentiometer and the base voltage with which it compares an input signal is 1 milli-volt, then the full scale voltage would be (10) (10) (1 mv) or 100 mv (20 dB on voltage = 10:1, so 40 dB is 10 × 10 or 100:1). On the same basis, if one apple were the reference, then 6 dB apples would be about 2 apples . . . 6 dB ≅ 20 log (2/1).

Vibration analyzers, regardless of what others may tell you, are simply voltage analyzers for the most part. The fun is learning how many volts are in a mil peak-to-peak, an inch/second or 1 g.

Table 3-1 is a chart to help you understand some multipliers in dB since they are definitely non-linear. I failed to mention that if the dB level is a negative number, then the ratio is reversed, e.g., −20 dB voltage gain would imply an output of 1 volt for an input of 10 volts. I find that it helps to remember about three of the dB conversions and because I rarely deal in power and almost totally in voltage, my three are: 20 dB is exactly 10:1; 10 dB is approximately 3:1, and 6 dB is approximately 2:1. Therefore, 46 dB would be (10) (10) (2) or about 200:1.

Table 3-1.

Pressure ratio	dB − + ← →	Pressure ratio	Pressure ratio	dB − + ← →	Pressure ratio
1.000	0.0	1.000	.316	10.0	3.16
.988	0.1	1.012	.251	12.0	3.98
.977	0.2	1.023	.199	14.0	5.01
.966	0.3	1.035	.158	16.0	6.31
.955	0.4	1.047	.126	18.0	7.94
.944	0.5	1.059	.100	20.0	10.00
.891	1.0	1.12	.0316	30.0	31.62
.841	1.5	1.19	.0100	40.0	100
.794	2.0	1.26	.0032	50.0	316
.708	3.0	1.41	10^{-3}	60.0	10^3
.631	4.0	1.58	10^{-4}	80.0	10^4
.562	5.0	1.78	10^{-5}	100	10^5
.501	6.0	2.00	10^{-6}	120	10^6
.447	7.0	2.24	10^{-7}	140	10^7
.398	8.0	2.51	10^{-8}	160	10^8
.355	9.0	2.82	10^{-9}	180	10^9

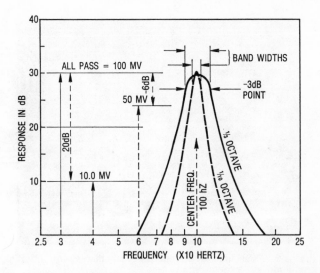

Fig 3-1—Vibration amplitude vs. frequency plot from spectrum analyzer and graphic level recorder.

46 dB = 20 dB + 20 dB + 6 dB $\cong$ (10) (10) (2). Recall that to multiply you simply add the logarithms.

In concluding the section on log scaling, it may be apparent that the reference value is important and should not be 0 but low enough to approximate 0 in reference to most values in terms of dB. On the conversion chart in Chapter 1, Fig. 1-4, some of these values were stated plus the power and pressure (voltage) relationship was listed. Now that the dB scaling has been reviewed, it is well to list some reference values in current use as taken from the *Endevco Dynamic Test Handbook*.[1] The subscript "o" could just as well be "r" for reference:

Sound power . . . $P_o = 1$ pW $= 10^{-12}$ W $= 10^{-5}$ erg/s

Airborne sound pressure . . . $P_{og} = 20\ \mu$ Pa $= 0.0002\ \mu$ bar $= 0.0002$ dyne/cm² $= 2.9 \times 10^{-9}$ psi

Waterborne sound pressure . . . $P_o = 1\ \mu$ Pa $= 10\ \mu$ bars $= 10^{-5}$ dyne/cm²

Acceleration . . . $a_o = 1\ \mu$ g (where g $= 9.80665$ m/sec.² $= 386.089$ in./sec.²)

Velocity . . . $v_o = 10^{-8}$ m/sec. $= 10^{-6}$ cm/sec.

Pa = pascal = N/m² = Newton/meter² $= 1.45 \times 10^{-4}$ psi

bar = 14.5 psi $= 10^5$ N/m² (Pa)

dyne/cm² = 0.100 N/m² (Pa)

erg/sec. $= 10^{-7}$ W (watts)

FILTERS

Unless an analyst wants a supreme headache, he will be using filters to condition a signal into the discrete frequencies for the purpose of analysis. The analyst needs filters as a radio operator needs a tuner.

A rotating machine is much like a section of the country with many radio stations. Each station is sending signals at its own particular frequency. If you want to hear some local radio station then the tuner, which is often called a hetrodyne resonance filter, must be tuned to a specific frequency, say 950 kc (kHz). If you want to investigate oil whirl on a rotating machine, then the 40-45 percent running speed frequency area could be investigated with a filter of some type.

Fortunately for us, band pass filters are rated in many ways and are of many types. The filter bandwidth is stated at −3 dB down from the center frequency peak response. Another reference of the filter characteristics is in terms of "Q" of the filter. This refers to the center frequency divided by the high end minus the low end frequency limits, generally at −3 dB. A "Q" of 10 tuned at 100 Hz would allow the signal to pass between frequencies of 95 Hz up to 105 Hz (100 Hz/10 Hz = 10 (Q)).

This same filter with a "Q" of 10, as a percentage filter may well be referred to as a 10 percent filter. The 10 percent may well be stated + 5% − 5%. Another reference may be the word "broad." A 5 percent filter may be classed as "sharp," high Q(20) or by octaves.

Filters with a fixed bandwidth may be great at high frequencies but poor at low frequencies. The constant percentage filters may be good at low frequencies and poor at high frequencies. For example, the 10 percent filter above would allow signals to pass from 9.5 kHz to 10.5 kHz when tuned in on a gear mesh frequency of 10 kHz (typical of a 60 tooth pinion at 10,000 rpm). The first four side bands would pass with the gear mesh frequency. This is a job for a real time analyzer with a resolution of 250 to 500 lines, but will be discussed later on analyzers. Band pass filters attenuate both low and high frequencies.

The trace recording shown in Fig. 3-1 was made on a General Radio 1911 analyzer and graphic level recorder. The signal was at 100 Hz (6,000 cpm) at a level of 100 mv RMS (282.8 mv peak-to-peak). The paper is log scaled with 40 dB full scale (span), i.e., it covers a range of 100:1 and the signal was attenuated externally to display at 30 dB up the paper. The narrow response was with a 1/10 octave band filter; the wider response is with a 1/3 octave band filter. (Note: one octave *up* from 100 Hz would be 200 Hz; *down* would be 50 Hz.)

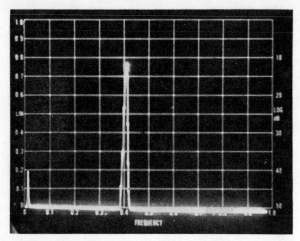

Fig. 3-2—Real time spectrum analyzer display of vibration amplitude vs. frequency.

The same signal was displayed on a Spectral Dynamics 330A analyzer as shown in the photo of Fig. 3-2. Sweep rate was to a 15,000 rpm full scale and with a bandwidth of 90 rpm or 1.5 Hz. The sharpness is easily noted.

These two analyzers just happened to be available to me at the time this article was being written. The analyzers are used for two different purposes and for those purposes the filters are fine.

A Bently Nevada low pass filter was then tuned to 6,000 cpm and about 71 percent of the signal strength was displayed. By tuning the low pass filter to 12,000 cpm full signal strength was obtained. The same operation was performed with the high pass filter and full strength came in below 3,000 cpm.

The low pass filter can be a fixed or tunable type. It is designed to pass low frequencies and attenuate high frequencies. A proximity probe could use a low pass filter to allow running speeds and certain multiples of running speeds to pass, but not higher frequencies from surface defects on the shaft. A low pass filter was advised in Chapter 2 to filter out an accelerometer's resonance and resulting high amplification. It should be noted that the roll-off characteristics of low pass (or high pass) filters might be at say 6 dB, 12 dB or 18 dB per octave. My suggestion when using these filters is to first check the output versus frequencies on a calibrated signal. Normally, if you tune the filter to twice the frequency of interest, all the signal will get through. The converse is true for high pass filters. Here the filter is designed to pass high frequencies and attenuate low frequencies.

Another filter that I have used quite a lot is the band reject filter. Sometimes it is called a notch filter. When band pass filters (percentage, fixed band, octave) are used to tune in the known frequencies, e.g., running speed frequencies of a machine, the question then arises as to the other frequencies (assuming one is not using a broad range RTA). Generally, if the "filter out" (all pass) vibration values agree close to the filtered value, then the signal should be purely synchronous with the tuned frequency. The band reject or notch filter is handy, if available, to select as a filter mode since it rejects the frequency tuned

and lets any other signals through. If no signal is noted, why waste time searching for other frequency(s).

One care in using any filter system is not to sweep the range so fast that you miss data. Back to the radio, if you sweep the tuner too fast then you miss your station. One manufacturer states that the time for a filter to reach 99 percent of its final value is $t = \dfrac{4}{BW @ -3 \, db}$ sec. The maximum sweep rate for a narrow band filter for less than 1 percent error due to filter blurring is $R = \dfrac{(BW)^2}{4}$ Hz/sec.

Some filter manufacturers will further describe the shape of their band pass filters by using a term called shape factor. This helps to define the degree towards a theoretical rectangular filter shape. The shape factor relates the filter bandwidth at -60 dB to that at -3 dB, e.g., a shape factor of 4 would imply a width at $1/1,000$ (-60 dB) of the peak of 4 times that at -3 dB or 0.707 of the peak.

There is one remaining passive filter that can cause you a great many problems because it isn't considered a filter but it is just that in actual practice. When you couple a signal to a measuring system it is not uncommon to use coupling capacitors. In the case of the accelerometer, you must couple the charge generating sensor to a measuring system. The cable, the sensor and other connections have capacitance. The measuring system generally has capacitance, particularly if AC coupling is incorporated, which means that a capacitor is used, often to block DC voltage. For example, the oscilloscope that you use may state that the input is across 47 pico-farads. Further, the instrumentation system has an input impedance, e.g., 1 meg ohms on most oscilloscopes. The point is that a circuit of capacitance and resistance is completed and there is a limiting low end frequency of this high pass network that may cause you a problem.

I can explain this best with an example. Suppose that I have an accelerometer with cable that has 70 mv/g voltage output with 1,000 pf of capacitance, and I connect that to an oscilloscope direct with a 1 meg ohm input impedance. What is the lowest frequency that will give me 70 mv/g output when the sensor is perturbated at 1 g? Up jumps another one of those relationships of electronics getting in the way of first class mechanical engineering work. The low end frequency is expressed by f (Hz) $= \dfrac{1}{2\pi (R \, ohms)(C \, farads)}$. In this case $f = \dfrac{1}{[2\pi (10^6)(10)^{-9}]}$

$= 159.15$ Hz. This means that if data were being taken at about 4,700 rpm (78 Hz), the output will probably be about -6 dB down or 30-35 mv. This could cause one to think the vibration level is $\frac{1}{2}$ g when in fact the measuring system is at fault. The insertion of a preamplifier in this circuit with a 2 gigaohm input impedance would alter the low end frequency to 0.16 Hz (10 rpm) and the full signal would come through. Recall the attenuation of a high pass filter could be on the order of 6 dB per octave or greater. In this example, data is being taken about one octave down (1:2) from the cutoff frequency of a high pass filter system. Take care of coupling capacitor sizing and input impedances of measuring systems to be purchased. (Note: the capacitor can be selected by swapping places with C and f above.) The same errors are being committed daily by taking critical voltage measurement with low impedance voltmeters.

A complete book could be written on filters and scaling, but the important fears and explanations have been touched on in this section. Three things happen when filters are being used:

1. Some information is generally being removed from the signal

2. Some attenuation of signal is generally affected

3. Some phase shifting is present.

Reflect for a moment on using a tunable vibration analyzer and an amplitude meter and phase display, e.g., strobe light. As you tune in the precise frequency of interest, e.g., running speeds, the amplitude increases and the phase is ever shifting and is only accurate exactly tuned at the running frequency.

LITERATURE CITED

[1] *Endevco Dynamic Test Handbook,* Rancho Viejo Road, San Juan Capistrano, Calif.

REFERENCES

General Radio Handbook Of Noise Measurement, Concord, Mass. Diehl, George M., *Machinery Acoustics,* John Wiley & Sons, Inc., 1973.

Phase and Basic Balancing

Phase measurements are generally the least performed and yet the most necessary in vibration studies. Phase measure is important when the problem is not too obvious. I might add that phase is the most difficult to explain for several reasons. First, it has no basic standardized terminology because people use the information in various areas and with varying types of instrumentation, i.e., each in his own area of study feels his phase measure and terminology are quite straightforward. Second, it is a measure of one change or event with respect to the sensor or to some timing mark. Third, most any resonance will change the phase. Nodal point shifts through a measurement probe change the phase and two phase changes are often in effect simultaneously.

It seems to me that phase does imply the measure of motion at one location with respect to a measure or motion at some other location. That relationship could be in degrees, time (seconds), or fractions of a revolution or cycle.

In Fig. 4-1, two signals of equal frequency are seen to be out of phase with one another, and the resulting waveform of the two signals is also shown. The upper signal *leads* the lower signal by 120°, or it could be said that the lower signal *lags* the upper signal by 120°. This leading or lagging could be referenced at the neutral axis or at the peaks (crests), but the phase referencing should be consistent.

In Fig. 4-2A, there are two signals of different frequencies e.g., frequency ratio of 2:1. While the signals are sinusoidal and stationary, any phase difference would need explanation because one waveform will repeat itself twice to one complete cycle of the other waveform.

The illustration could easily be that associated with an oil whirl frequency which may have locked onto a rotor critical frequency near one half and coupled into a fixed half frequency. Fig. 4-2B is an actual signal taken from a turbine with a loose bearing support dropping the shaft critical near the oil whirl frequency. A periodic one-half running speed frequency exists, not typical of pure oil whirl. In my experiences I have found that a pure oil whirl will generally occur at about 42-44 percent of running speed frequency. The pattern is typified by Fig. 4-2C.

Oil whirl occurs generally at medium to high speeds when the bearings are lightly loaded. The journal (shaft) is traveling at rotor speed and the bearing is hopefully traveling at zero speed. The oil, which is being sheared continually, is traveling at less than one-half speed due to slippage. The oil wants to travel at one-half speed and with different surface conditions, concentricity, and clearance, it may well approach one-half speed (rotor). Some people claim that it can exceed one-half rotor speed.

WAVE FORM #1 — 1.5 SIN ($\omega t + 60°$)

WAVE FORM #2 — 2.5 SIN ($\omega t -60°$)

2.18 SIN ($\omega t -23°$)

COMPOSITE WAVEFORM

Fig. 4-1—Addition of two sine waves of equal frequencies.

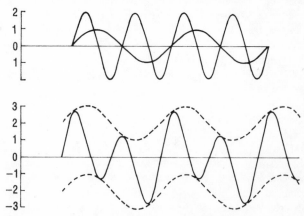

Fig. 4-2A—Addition of two sine waves with frequency ratio of 2:1.

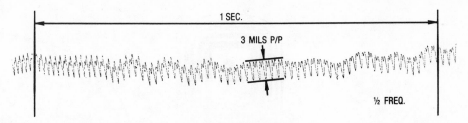

Fig. 4-2B—Hand-held vibrograph of turbine with loose bearing saddle support. Note the one-half running speed frequency.

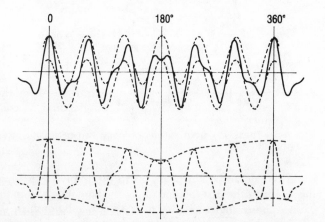

Fig. 4-2C—Typical oil whirl waveform where frequency of one waveform is nearly one-half of the other waveform.

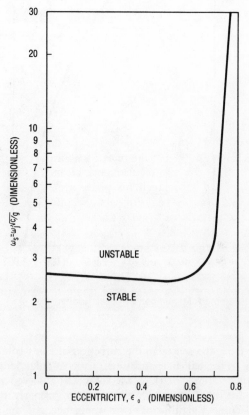

Fig. 4-3—Stability for the unloaded short journal bearing.

To reference John Sohre for a moment as I have always appreciated his description of oil whirl—if the bearing and, hence, the oil in the bearing is lightly loaded, the oil carries the shaft about in the bearing much like a surfboard which is caught in a wave. If, however, the shaft is heavily loaded, e.g., as a battleship, it is not sensitive to oil film forces. It seems obvious that forward precession is a characteristic of oil whirl. This forward precession implies that the vibration signal *precesses* in the same direction as the journal or shaft.

Dr E.J. Gunter, Jr.,[2] has given a guideline to the threshold of instability (ω_s) for straight sleeve bearings with length to diameter ratios of 0.5 to 1 (short bearing theory). If $\omega_s > 2.5$ as shown in Fig. 4-3, instability occurs at eccentricity ratios less than approximately 0.6 to 0.7. The ω_s is equal to the shaft or journal speed in rad/ sec times the square root of the ratio of radial bearing clearance in inches to the gravitational constant $g = 386$ inches/second. This limit assumes a rigid rotor, as instability could occur at lower levels with a flexible rotor. The ω_s would be less as k_{shaft} is less (refer to Chapter 1). Fig. 4-4 can be used to find the eccentricity value required for stability determination.

Back to phase again, another method of indicating phase is by the use of a strobe light which is triggered to flash on some signal generated by the sensor analyzer as it follows a vibration pattern. Often, the strobe flashes as the signal goes from negative-to-positive or positive-to-negative. Vibration analyzers that I have used do exactly that. The analyzer builder may send you a chart showing phase angle from the above crossover point to the next positive peak. As you will see later, this convention of

trigger-to-the-next-positive-peak has been the standard that I have seen on other systems where a trigger (timing mark or clock start signal) references to the next positive vibration signal from a sensor.

Of more importance to me is the phase angle between a sensor's position and the distance that a high spot or heavy spot has traveled at the time the strobe light freezes some reference mark on, say a rotor's shaft end or balance ring. With a tunable analyzer, this phase can vary with the filter tuning, i.e. correct only at center band of the filter. It can also vary with the phase shifting of a system at low frequency, e.g., below roll-off frequency. Both of these faults were discussed in Chapter 3 under filters.

Fig. 4-5 shows the change of this phase angle with one analyzer in the "displacement" mode (integrating the velocity signal causes a 90° phase shift) expressed here as phase lag from the sensor (seismic-velocity) to the heavy spot, below the first rigid rotor critical. This plot of lag would be very helpful in balancing at, say 1,200 rpm as the phase lag would indicate that the heavy spot would

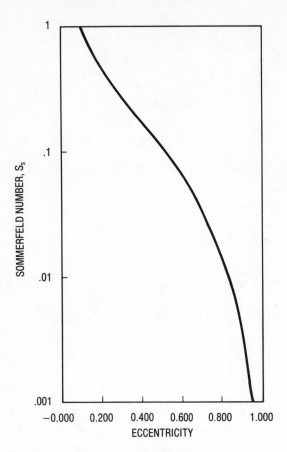

Fig. 4-4—Sommerfeld number vs. eccentricity.

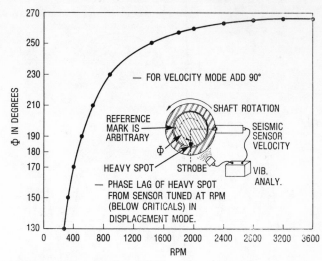

Fig. 4-5—Phase lag from seismic-velocity sensor to the rotor heavy spot.

be 244° from the sensor in the direction of shaft rotation. First, the rotor will be positioned to the angular location where the reference mark on the rotor was observed when running at 1,200 rpm with the analyzer properly tuned to 1,200 cpm. A significant error would occur if the analyzer was not in tune or if the speed changed prior to recording the phase angle without retuning the analyzer to the new speed. **Caution:** The curve as plotted would be in error if the analyzer was in the "velocity" mode. However, one could *add* 90 degrees to the values determined which should put the lag *near* zero degrees in "velocity" for a velocity sensor above the roll-off frequency (approximately 3,000 rpm) which is as it should be. A trial weight could be placed in a position 180 degrees from the heavy spot for the first trial in balancing, determining an influence coefficient from this trial weight.

Phase measures can also be used to determine whether a foundation plate or section of piping is in phase with a running machine. One technique is to place the sensor on the baseplate, for example, and let the strobe light indicate on the running machine. The order of resonance, if any, is also confirmed. If three reference marks were indicated on the shaft, the baseplate would be vibrating at three times the frequency of the rotor and in resonance (excited by) with the rotor. Conversely, the sensor could have been placed on the rotor, tuned to running speed and the strobe light played about the area looking for anything that could be "frozen" by the light. If so, that item

could be said to be "in phase" with the rotor. By moving the sensor in the first method along the baseplate and indicating phase angles on the rotor, phase relationships 0, 90, 180, 270 . . . can be determined along with nodal points of the baseplates.

I am always amazed when two sensors are placed on each end of a machine, e.g., one at the top of the steam end bearing and the other at the top of the exhaust end bearing of a steam turbine operating above the first rigid critical in a conical (pivotal) mode and "phase" measurements made between the two sensors. The fact that the signals are out-of-phase by 180 degrees often has some people scurring about looking for strange, bad contributors, such as misalignment, when the phase relationships are in complete agreement with the rotor mode shape.

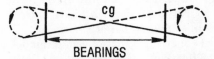

Switching to a displacement sensor for a moment, only for simplicity, suppose a proximity sensor was placed at the 12 o'clock position or on top of the steam end bearing and the rotor was turning counterclockwise as viewed from the driving or governor end of the turbine. Let us also place another probe near the same position at the same angular location and let that probe look at the shaft which now has a small key slot, say ½ inch wide x ¾ inch long x 60 mils deep. As this key slot passes the probe, a spike in the signal will be included since, as

the slot is first seen by the probe, it indicates the shaft is further away, thus, an increase in voltage output by the sensor. As the trailing side of the slot passes, the shaft distance to the probe returns and the voltage reduces to the level as seen before the slot appeared.

The signal from the vibration sensor will be put on time trace of an oscilloscope, and the slot-probe produc-

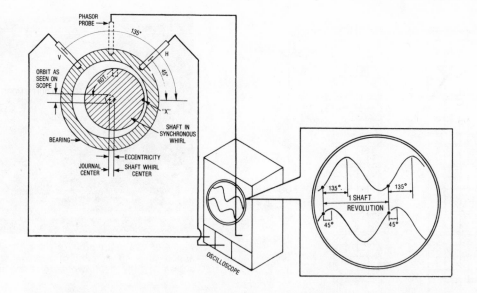

Fig. 4-6—Diagram showing phase marking by using blanking marks on an oscilloscope display.

ing signal will be used to synchronize the trace and AC coupled to the "Z" input or "blanking" input to the CRT. Recall from Chapter 3 that AC coupling means putting a capacitor in series with the signal. "Blanking" in scope lingo means voltage changes which produce intensity changes in the display. An increase in negative voltage causes the electron flow to increase and the display is brighter. Voltage changes in the other direction cause the display to dim. To further confuse the poor mechanical people using the scope, the manufacturers of scopes use the "cathode" of the CRT *or* the "grid" for blanking the signals. In one case, what I have just explained is true—in the other case, it is exactly reversed. (Using the capacitor to AC couple causes the intensity change to be limited to the "instantaneous" effect of the slot or phasor pulse.)

On Fig. 4-6, the trace is seen from such an arrangement. The angle separating the blanking mark (occurs when slot passes probe) on the trace and the next positive peak of the signal is commonly referred to as a phase angle. Another way of stating the condition of the rotor using this phase angle may help. Forty-five degrees after the slot passes the phasor probe, the high spot of the rotor passes the vibration probe. Getting still a little more practical, one could align the slot on the rotor (by strap rotation of the shaft at rest) to coincide with the phasor probe, then starting at the vibration sensor (in this case it is at the same location), mark off 45 degrees from the sensor against rotation and place a mark (x). This mark (x) represents the high spot of the rotor during shaft whirling at rotor speed. The "high spot" can be thought of as the part of the rotor that comes the closest to the probe when the rotor is operating at the speed the data was taken. Recall from Chapter 2, that as the gap decreased, the negative voltage decreased (same as positive voltage increase) and a plus signal goes *up* on an oscilloscope for the vertical amplifier (*plus* goes *right* on the scope for a *horizontal* amplifier). The most positive or "up" position of the scope is the next positive peak.

If one were to balance the rotor, a trial weight placed 180 degrees opposite the mark (x) on the rotor should make the correction, provided the right trial weight is used. If too large a trial weight is used, the phase angle will reverse on spin-up, reading about 225 degrees (45+180). If too small a trial weight is used, the same angle appears; but the amount of vibration is less. These two logics are basic in understanding any balancing procedure. If neither option presents itself, it is apparent that single plane balancing will not work and a more complicated two plane, multiplane, or modal plane balancing method must be used.

One interesting aspect is present in the direct method just described. There is no phase lag present in this system, i.e., no time lags from the rotor-to-sensor-to-scope. However, you must realize that the heavy spot (point of mass unbalance eccentricity) need not coincide with the high spot at the balance speed. One advantage is that the phase angle with hold for changes in speed, provided a resonance (critical speed) is not occurring.

LISSAJOUS OR ORBIT MEASURE

If two signals at the same frequency but with different phase relationships are displayed relative to one another on an oscilloscope (no time base used at all) a lissajous pattern is produced. Fig. 4-7 illustrates the possible phase relationship between the two signals. This is nothing new and this chart was obtained from Radar Electronics Fundamentals[3] and was also shown in a previous article[4] by the author. The signals need not be at the same frequency. The oscilloscope photo in Fig. 4-8 was taken of oil whirl on a rotor model and shows 48 percent running frequency plus running frequency. Note "blanking" mark on each shaft revolution. Internal loops indicate "forward" precession, e.g., oil whirl. External loops indicate "reverse" precession, e.g., friction whirl. The number of internal loops plus one (n + 1) indicates the frequency ratio. The

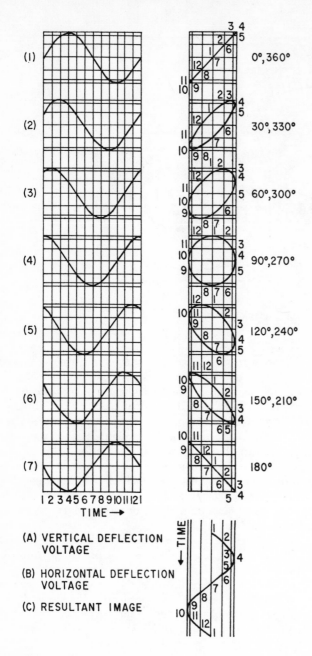

Fig. 4-7—Lissajous phase relations.

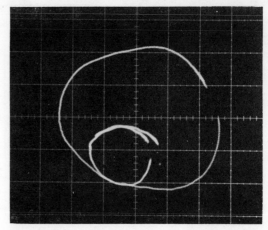

Fig. 4-8—Oil whirl orbit of rotor showing running speed plus 48 percent of running speed. Photo shows slightly over two revolutions of the shaft.

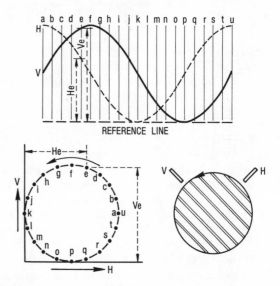

Fig. 4-9—Development of an orbit pattern from simultaneous vertical and horizontal waveforms.

number of internal loops minus one $(n - 1)$ indicates the frequency ratio. (Exception: 2:1.)

Fig. 4-9 illustrates how a lissajous or orbit pattern is developed. One can simply make their own if the two signals vertical and horizontal are known and recorded simultaneously. It is important to point out that the horizontal sensors should always be placed 90 degrees to the right of the vertical sensors, regardless of the shaft rotation. This reverts back to plus "up" and plus "right" logic, also typical of scopes. The vertical probe need not be in the true vertical position; but one should keep in mind that as a shaft, for example, moves towards the vertical probe, motion will move straight up on the scope trace.

BALANCING

There are several techniques for single plane balancing rotating machinery of use to the practical engineer. Balancing is one operation that puts the phase measure to very good use.

First, assume that a single mass disc or impeller is to be balanced, using the influence coefficient technique and vector solution on polar paper (see Fig. 4-10). For practical illustration, we will use a typical field balancer/analyzer with a strobe light to read phase. A seismic-velocity sensor is placed on the nearest bearing, and the strobe light displayed on the coupling end of a machine with the rotation counterclockwise as one views the coupling end. Speed will be 6,000 rpm and a rotor weight of 1,000 lbs.

The analyzer is tuned to exactly 6,000 cpm (100 hz), and a reading is taken as 3 mils p/p at a phase angle of

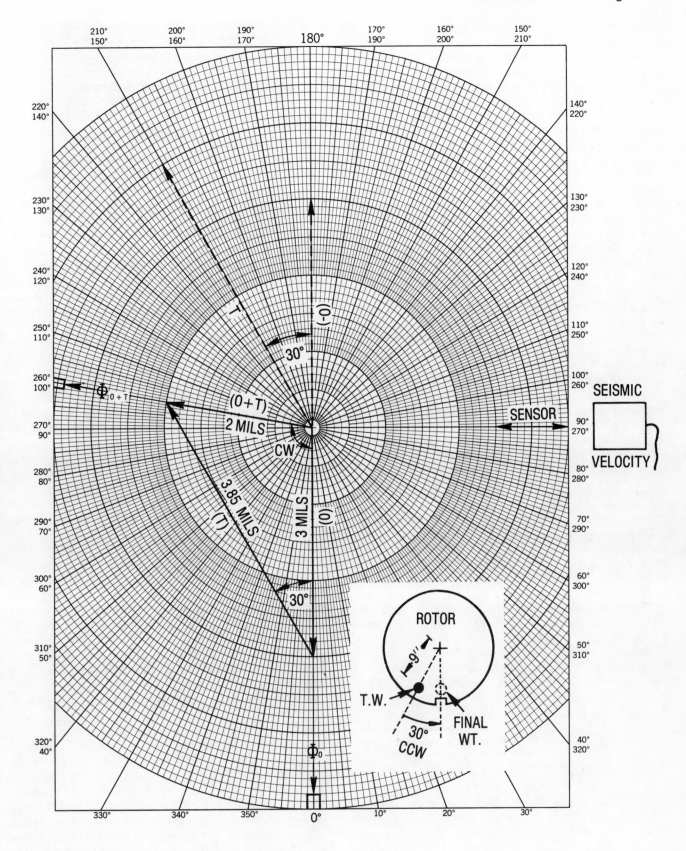

Fig. 4-10—Single plane balancing using graphical vector solution.

180 degrees, using the vertical position (12 o'clock) as zero degrees. The sensor is located at 90 degrees (3 o'clock). This original reading will be called "0" and plotted on polar paper as a vector 3 inches long (1 inch/mil) and at a phase of 180 degrees. Now the intent from this step forward will be to obtain a response from a trial weight (TW) and create a new vector equal to minus 0 (− 0); or from our first vector 0 to a new vector equal to 0 (3 mils), but at 0 degrees rather than 180 degrees as first recorded.

The phase angle of the reference mark is really arbitrary at this point, which confuses a lot of people in balancing. It is a reference to start measuring changes of phase with trial weight placement. The reference mark is often a keyway (if you have more than one keyway, be sure you know which one is being used). The phase angle is also a means of returning the rotor to the same position where information on placement of correction weights is determined.

Step 2 calls for determining a trial weight and placing it on the rotor. I always use a trial weight, causing a force equal to 10 percent of the static weight of the rotor at rest; but determined at the actual running or balancing speed. Since the force (F_{tw}) of the trial weight equals the mass of the rotor (M) times the acceleration (A) caused by the eccentricity of the trial weight at the rotational speed, we are back to Newton's F = MA. In this case set F = 1/10 (1,000 pounds) = 100 pounds. To simplify the MA term, which you may recall as W/g ($\omega^2 r$), by rearranging units, one ends up with F = 1.77 $(rpm/1,000)^2$ (ounce-inches of trial weight). So, ounce-inches = $\dfrac{100}{1.77\ (36)}$ = 1.57 ounce-inches or 0.17 ounce at 9-inch radius, which looks about right for this example.

If we knew the phase lag of the analyzer and the sensitivity of the rotor to unbalance, we could try for a "one shot" correction at this point. However, we will place the trial weight at 30 degrees clockwise from the reference mark (keyway).

Step 3 calls for running the rotor back to 6,000 rpm and recording the amount of vibration and the new phase angle. This new reading is 2 mils p/p at a phase angle of 260 degrees. This vector is labeled original plus trial weight (O + T) and is plotted on the polar paper. If one would now subtract the O vector from the O + T vector, this should yield the T vector (O + T-O = T). This will also keep the arrow tips of the vectors straight because O + T should end up (O + T), i.e., vectors with heads-to-tails always *add*; heads-to-heads indicate *subtraction*.

Step 4 calls for ratioing the length of T vector (measures 3.85 mils) to that of the O vector in order to modify T to be the length of O. This ratio is simply the O/T multiplier of (TW) trial weight. O/T is 3/3.85 times 0.17 ounce or 0.132 ounce.

Step 5 calls for determining the shift of the trial weight to affect a shift of the T vector to the −O phase position. The angle formed by translating the T vector to the center of the polar plot is also the same angle as between vector O and vector T. This angle is 30 degrees, and the shift would be from the present position at 30 degrees by −30 degrees to 0 degrees. The direction of shift is not always the same. The angle between O and T is always correct, but the shift in that direction will always be opposite the direction that O + T moved from the original O vector. The O + T is CW from O, so the TW shift is CCW by 30 degrees. One should prove this out to satisfy themselves, e.g., place a weight on a rotor; note the phase angle and shift the weight 30 degrees to the right. The phase angle will shift 30 degrees to the left.

Step 6 would call for placing the new trial weight 0.132 ounce at 9-inch radius and at an angle on the rotor equal to 0 degrees from the reference mark (keyway). The answer obtained is typical of corrections where keyways have not been compensated by balancing.

Accuracy of determining the angles, and weighing and placing the trial weights cannot be overemphasized. **Do sloppy work . . . get sloppy results.** The use of 10 and 20-inch maneuvering charts for plotting will improve accuracy.

Using a calculator will improve the accuracy over graphical plotting, but is difficult to cover in a primer. However, the steps as taken from the format taught at the University of Virginia by Dr. E.J. Gunter, Jr., are shown in Fig. 4-11, but not discussed. One versed in polar and rectangular coordinates will see the extreme similarity to the graphical vector solution just performed. I would note that a right hand coordinate angle system is used, phase angles are taken in the direction of rotation, and the trial weight positioning on the rotor are always against rotation from the reference mark (keyway).

Reading of the actual phase angles with a strobe light can be enhanced by cutting a shaft's diameter (plus) out of a polar sheet and placing it on the machine with tape for temporary reading. Darkening the area or taking the data at night also helps, and it rids you of many unnecessary bystanders. A portable weighing scale, typical of O'Haus, is also very helpful. Buy the extra weights also for extending the range of the weighing scale. Do not use the scale in a windy area. Finally, it is not necessary to add weight. Removing weight is much the preferred method. A friend of mine once remarked, "I ain't thrown a hole off yet." Remember to shift your location by 180° if you solved your balance problem on a "weight-add" basis.

Knowing the phase lag of an analyzer on a particular machine, e.g., take our graphical example and phase lag from Fig. 4-5, one could have placed the trial weight, for that unit only, at 175 degrees from reference mark (CW) or 265 degrees phase lag from the sensor in the direction of rotation right off the bat and perhaps made a successful balance on the first trial. If one will record the unbalance corrected per ounce-inch or gram-inch along with the apparent phase lag of the heavy spot (180 degrees from final placement of correction weight) from the sensor, that machine (with that same analyzer) should allow a one shot correction in the future. The

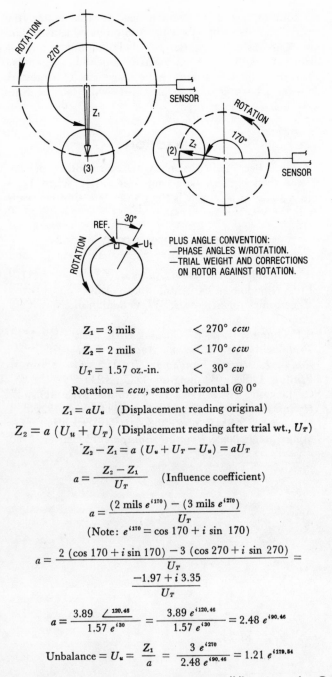

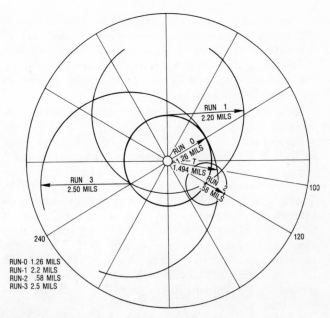

Fig. 4-12—Four-run balancing method.

$Z_1 = 3$ mils $\qquad < 270°\ ccw$

$Z_2 = 2$ mils $\qquad < 170°\ ccw$

$U_T = 1.57$ oz.-in. $\qquad < 30°\ cw$

Rotation $= ccw$, sensor horizontal @ $0°$

$Z_1 = aU_u$ (Displacement reading original)

$Z_2 = a\,(U_u + U_T)$ (Displacement reading after trial wt., U_T)

$Z_2 - Z_1 = a\,(U_u + U_T - U_u) = aU_T$

$a = \dfrac{Z_2 - Z_1}{U_T}$ (Influence coefficient)

$a = \dfrac{(2\ \text{mils}\ e^{i170}) - (3\ \text{mils}\ e^{i270})}{U_T}$

(Note: $e^{i170} = \cos 170 + i \sin 170$)

$a = \dfrac{2\,(\cos 170 + i \sin 170) - 3\,(\cos 270 + i \sin 270)}{U_T} = \dfrac{-1.97 + i\,3.35}{U_T}$

$a = \dfrac{3.89\ \angle\ 120.46}{1.57\ e^{i30}} = \dfrac{3.89\ e^{i120.46}}{1.57\ e^{i30}} = 2.48\ e^{i90.46}$

Unbalance $= U_u = \dfrac{Z_1}{a} = \dfrac{3\ e^{i270}}{2.48\ e^{i90.46}} = 1.21\ e^{i179.54}$

Correction weight $U_B = -U_u = 1.21$ oz.-in. $e^{-i0.46} = 1.21$ oz.-in. @ 0.46 degrees with rotation from keymark.

Fig. 4-11—Single plane balancing using mathematical solution.

original heavy spot in our example was 180 degrees opposite the keyway (reference mark) or about 270 degrees past the sensor.

Balancing without phase measure. It is possible to get extreme accuracy in balancing by determining the correct balance position using nothing but a simple vibration meter. This requires three runs of the equipment, if successful. It is referred to as the Four Run Method. I learned this technique from Michael Blake, consultant,

when he worked for my company in the early 1960s. The technique is also covered in the *Vibration and Acoustic Measurement Handbook* by Blake and Mitchell, available through The Vibration Institute and illustrated in Fig. 4-12.

This procedure is good for things like cooling tower fans where the fan speed is different from that of the motor drive and getting phase data is either expensive or unhealthy. It is also good when phase data are inconsistent, such as two blowers with motors using the same platform or tower to be supported, thus, giving a beat frequency from each machine's feedback with slightly different slip frequencies on each drive, i.e., slightly different operating speeds. This causes a strobe light to oscillate about without clear definition of "frozen angle."

First, the machine is running, and the sensor for the amount of vibration is placed generally against the nearest bearing. Record the amount of vibration. This is the original (O) reading (1.26 mils in example). Draw a circle to some scale with the radius equal to the O reading (1 inch = 1 mil in example). Stop the machine. From some place on the element to be balanced, lay out a circle on the rotor or at least identify the rotor in some form of angular (360 degrees) layout for three trial weight placements at the same radius, but three separate locations on the rotor using the same trial weight.

Assume the plot taken was for a unit with six vanes. Start with vane one, which could be marked on the rotor. Place a trial weight at vane one. The trial weight in the example was 8 gms. Run the machine at the same speed as before and record the amount of vibration. Plot Run 1 value (2.2 mils), using zero degrees position on the O drawn circle as a center.

Shut down the machine, remove the trial weight from vane one position, and move to vane three position, i.e., 120 degrees for the trial weight.

Run the machine again and record Run 2 amount (0.58 mils). Using 120 degrees on your plot and its intercept on the O circle, draw another circle with Run 2 amount as the scaled radius (0.58 inches). Shut down the machine and relocate the trial weight (same) to vane five (240 degrees).

Run the machine again and record the amount of vibration (2.5 mils). Plot a fourth circle, using 240 degrees and the O circle as the center with a radius equal to Run 3. Locate the area where the Runs 1, 2, and 3 circles intercept. Locate a center of that small area. From that center, connect a line to the center of the O circle. The length of that line (1.494 inches) labels and measures as the T vector (1.494 mils). Ratio the O amount to the T amount as before (1.26/1.494) and multiply that value against the trial weight used. This is the correction amount (6.746 gms) and the placement will be at the same radius and at a point on the rotor described by T's position. In this case, 100° clockwise from vane one.

The run plotted was actual data, i.e., final value was 0.1 mils. If the solution position along T is not practical on the rotor, e.g., between two blades, vectorially solve the amount in each blade, using the calculated amount as a resultant vector at the T position angle.

Balancing using the orbit. This procedure is well-covered in Chapter 11. Current instrumentation, however, greatly simplifies the correct angular position. The orbi-tal concept, while it definitely has advantages on trial weight placement in balancing, also offers benefits in failure diagnosis, e.g., location of blade fouling or missing blade. As outlined in an ASME paper[5] in September 1974, the orbit was used to guide placement of a pressure dam on a bearing to limit oil whirl, while lobe bearings were being manufactured.

CONCLUSION

I have been guilty along with many other persons of not taking phase data more regularly. When I get into a sticky problem, it is the phase data that generally leads the way out to success.

LITERATURE CITED

[1] R.G. Manley; *Waveform Analysis;* Chapman & Hall, Ltd.; London, 1950; Malmstadt, Enke and Toren; "Electronics for Scientists," W.A. Benjamin; New York, 1963.
[2] E.J. Gunter, Jr. & R. Gordon Kirk, Nasa; CR-1549; *Transcient Journal Bearing Analysis;* June 1970.
[3] *Radar Electronic Fundamentals;* U.S. Government Printing Office; Washington, D.C.; 1944.
[4] C. Jackson; "New Look at Vibration Measurement"; *Hydrocarbon Processing;* January 1969; Vol. 48, No. 1.
[5] C. Jackson; ASME 70-PET-30; *Using the Orbit (Lissajous) to Balance Rotating Equipment;* Denver, Colo.

Taking Field Data

Taking data in the field would seem to be a simple task. It is not that simple, but with some experience it will become second nature. One should plan a measurement either formally or informally.

In vibration analysis, frequency is always the key to the vibration source. The amplitude represents the magnitude of any forced function. The phase may determine position, the sequence, the modal shape, or it may determine one of several sources which transmit similar frequencies.

Consider the equipment to be measured. Are shafts exposed? Are bearings recessed? Must coupling guards be removed? Can a strobe light be used for phase measurement? Is a keyway available for phase reference? Can the equipment be stopped and restarted? If so, possibly reflective tape can be used for phase reference. If tape and photosensors are to be used, is a magnetic clamp or tooling necessary? If the bases are nonmagnetic, for example, aluminum or stainless steel, perhaps clamps are needed. If a motor is a driver, a probe may be needed to reach past the end bell and contact the bearing housing near the point where the grease fitting is located. The end bell should not be used, as it may amplify or have its own resonance. At what speed is the unit to be measured? What type of sensors are to be used? What temperature is to be encountered?

Anticipate problems to be analyzed. Be familiar with the machine internals before performing an analysis. If the manufacturer does not have an accurate cross-section drawing, 15 minutes with a maintenance man or manufacturer's serviceman is invaluable. It is unethical, in my

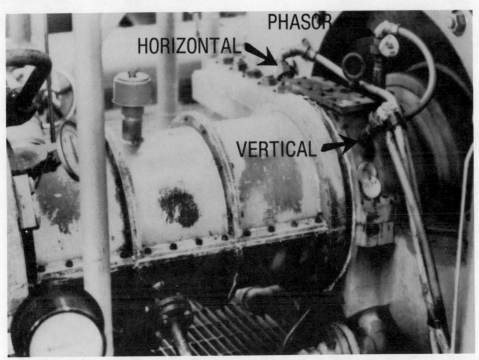

Fig. 5-1—Properly installed vibration sensors greatly simplify taking field data. Horizontal, vertical and phasor marking sensors are shown here on a turbine exhaust bearing.

opinion, for a manufacturer to submit so-called "typical" sectional drawings which do not correctly depict details of a machine built and sold to a customer.

Let's take an example. Suppose you are to investigate a steam-turbine driven centrifugal compressor with a gear increaser. It would be a good idea to make a list of the following information:

The turbine governor step-down (up) gear ratio(s), the number of teeth on the pinion and gear, operating speed, critical speed, number of nozzles, number of blades, type of coupling, type of bearings (radial and thrust), main gear teeth and ratio, gear-driven oil pump ratio and number of pump gears or elements, compressor stages and blading, number of balls or rollers on any antifriction bearings, type of seals (i.e., carbon, labyrinths, etc.), and type of assemblies (i.e., keyed, shrunk, sleeved, etc.).

Review the symptoms. There is a basic series of suspected problems based on the frequencies analyzed. If all the data are at running speed frequencies, then unbalance, bad journal bearings, fouling, loose elements, cracked blading, loose coupling, poorly fitted keys, and a whole list of *fundamental running speed* symptoms are possible.

Other causes can be removed from the list, such as oil whirl; hypo- or hypersynchronous resonances are out; case-to-case misalignment is improbable; gear problems are out.

Axial vibration data could be checked at this point which may define the motion as radial, if axial vibration levels are zero or very low. (See Fig. 5-1.)

Taking phase data at two separate, but related points on each end of the casing, will help define modal motion.

There is no substitute for inquisitive investigation with an analyzer. For example, if axial vibrations are high on the turbine exhaust and also on the gear input bearing, several problems are indicated. First, the coupling is not decoupling (locked). Second, if the gear's axial vibration value is higher than the turbine, it implies that the primary source is from the gear's direction. If the compressor's axial vibration value is low, then concentrate on the gear. Already you could expect thrust bearing problems on the gear or pitch line runout of the gear or pinion or both.

A continuation of symptom examples would be endless. You can recognize the analogy of a physician obtaining vital signs or data to narrow his diagnosis quickly to the problem area for correction.

I believe the greatest tabulation of symptoms was introduced by John S. Sohre.[1] Table 5-1 on pages 24 and 25 is adapted from the first of four charts with some comments taken from his fourth chart. Table 5-1 illustrates that vibration analysis is no quick, clean-cut, absolute solution. It can be, but often is not. The data under each of the categories are percentages of possibilities based on Mr. Sohre's experience. For example, under bearing eccentricity there is an 80 percent probability that this will occur at running frequency, and a 20 percent possibility it may be indicated at twice running frequency (2 x R F).

I would encourage one to use an oscilloscope to observe critical vibration patterns. Many symptoms are not periodic, continuous or consistent.

A loose thrust collar may vacillate from very smooth at one moment to very rough at another. Often the phase angle will jitter or move slightly, but never very far, i.e., it will often shift no more than 30° and return. The looseness is obvious, but the *confinement* of movement is also obvious.

A vibration meter is very nonchalant about this whole nervous affair. Meters are great averagers of events, whether in amount or frequencies. A frequency meter may indicate a predominate frequency of 30,000 cpm when actually two disturbances are occurring at 24,000 cpm and 36,000 cpm. Pay particular attention to the steadiness (or unsteadiness) of amplitude, frequency or phase indicators.

Tips on taking vibration readings. Taking direct readings from a shaft is often necessary. A vee-stick or "fishtail" of pine or Micarta makes a good shaft rider which can be placed in a shaft step-down. It helps to check a rotating shaft with a strobe light before using a shaft stick to be sure there are no shaft keyways, keys or other discontinuities in the measurement area. If a strobe is not available, touch the shaft with a sheet of cardboard. If the shaft is rusty or simply dirty, burnish the area with a wooden shingle on suitable thin stock. Polish the area carefully with a 12-inch to 18-inch strip of emery cloth followed by a few drops of lube oil. This effort will give very improved results. For safety, place the shaft stick on the shaft with the shaft surface rotating away from your body while taking the data. (See Fig. 5-2.) Permanently mounted proximity sensors within the bearing region, i.e., in a clean, smooth, oil-covered shaft area, would eliminate all the above frustrations.

Velocity sensors have often given false resonance on bearing housings from using either hand-held probes or light-duty vice grips. I prefer to clean the area with a rag or safety solvent and use a two-bar magnetic holder. Often, some of the paint should be scraped off for better contact.

Fig. 5-2—Taking shaft data after proper preparation using a shaft stick (fishtail) and a hand held vibrograph.

Mark the spot and repeat future readings at the same location for proper trends. (See Fig. 5-3.)

If threaded studs are used for velocity sensors or accelerometers, do not over tighten, and use studs *short enough* for contact of the sensor to the bearing without "free standing" on the stud, i.e., not like a mud ball on a fishing pole!

It is usually inconvenient to carry a large amount of equipment to the job site for several reasons:

- Weight limitations,

- Need for many data points simultaneously,

- Damage to the analyzers from shipment or environment,

- Need for permanent or comparative documentation,

- Analyzers in use at other locations,

- Transient data is spasmodic,

- More data is needed than analyzer's channel capabilities,

- Different types of analysis is needed requiring repetitive playback of data.

The tape recorder is a very versatile tool to record field data. One can bring the tape back to a lab or an office area where the data can be transcribed into different formats. Learning to use a tape recorder is somewhat difficult. I suggest that before purchasing expensive FM tape recorders, one should buy a high-quality AM tape recorder and learn some of the basic fundamentals from it. We have used AM tape recorders since about 1969 and have found them to be very helpful for several reasons. First, the AM tape recorder uses standard tape which is available at most music stores, or electronics supply houses. Second, and probably more important, an AM tape recorder will generally accept 25 volts or better without saturating the tape heads. In this manner, the data can be taken from the smallest of signals from an accelerometer on up to the strongest of signals from a velocity pickup or, for that matter, the data taken from proximity probes which often carry a bias DC voltage of seven or eight volts. Seven or eight volts on a FM tape recorder would saturate the tape recorder, and all the data would be lost. Most FM tape recorders can only accept 1 to 1½ volts RMS before saturation. Note: An RMS volt is equivalent to almost three volts peak-to-peak.

The first AM tape recorders that I had experience with were two-channel recorders. Later, we purchased an AM four-channel tape recorder. With the help of one of the instrument builders, Bently Nevada Corporation, the calibration pots were converted to precision pots and the two-pole phono tip connections were converted to BNC. The result was that we had a very usable system. Data are fed directly from the sensor straight into the tape recorder. The input impedance is about 150 K ohms.

Fig. 5-3—Two seismic-velocity sensors (vertical and horizontal) are shown held in place by two-bar magnets on a turbine exhaust bearing.

We usually use an oscilloscope to look at the raw data; and a channel of the oscilloscope can often be used to look at the playback data, as it is being recorded, just to be sure that the data are actually being recorded. It is important to remember that the playback data, as the data are recorded, are not necessarily equivalent to the data when played back in a playback mode. On a good recorder, it will be very close. Fortunately, the reproduce or playback heads follow the record heads in most AM tape recorders. For this reason, the data can be reviewed as the recorder is in the record mode; however, you should precalibrate the tape recorder for assurance that whatever signal is repetitive on playback, for instance, one volt peak-to-peak at 100 cycles per second. Of course, there is no control of the frequency. This becomes a quality condition of the tape recorder builder's playback speed and tape drive speed when on record. If the tape speed vacillates, naturally the frequency is going to vacillate in the same order. Any phase data may shift on playback from original data.

(Text continued on page 27)

Table 5-1—Vibration Analysis Symptoms.[1]

Cause of vibration	0-40%	40-50%	50-100%	1x running frequency	2x RF	Higher multiples	½ RF	¼ RF	Lower multiples	Odd frequency	Very high frequency
						Predominant Frequencies					
1. Initial unbalance	..	..	..	90	5	5	..	..	..	..	..
2. Permanent bow or lost rotor parts (vanes)	..	..	..	90	5	5	..	..	..	..	..
3. Temporary rotor bow	..	..	..	90	5	5	..	..	..	..	..
4. Casing distortion { Temporary	←——10——→			80	5	5	..	..	..	..	..
Permanent	←——10——→			80	5	5	..	..	..	..	..
5. Foundation distortion	..	20	..	50	20	..	..	..	..	10	..
6. Seal rub	10	10	10	20	10	10	..	..	10	10	10
7. Rotor rub, axial	←——20——→			30	10	10	..	..	10	10	10
8. Misalignment	..	..	..	40	50	10	..	..	..	..	..
9. Piping forces	..	..	..	40	50	10	..	..	..	..	..
10. Journal & bearing eccentricity	..	..	..	80	20	..	..	..	..	..	..
11. Bearing damage	20———————→			40	20	..	..	..	..	..	20
12. Bearing & support excited vibration (oil whirls, etc.)	←—10—→	←—70—→		..	..	..	10	10	..	..	..
13. Unequal bearing stiffness horizontal-vertical	..	..	..	..	80	20	..	..	..	..	..
14. Thrust bearing damage	90———————————————→						..	..	..	..	10
Insufficient tightness in assembly of:	Predominant frequency will show at lowest critical or resonant frequency						..	..	..		
15. Rotor (shrink fits)	40	40	10	..	..	..	..	..	..	10	
16. Bearing liner	90——→		..	..	..	..	..	..	..	10	
17. Bearing cases	90——→		..	..	..	..	..	..	..	10	
18. Casing & support	50——→		..	..	..	..	..	..	..	50	
19. Gear inaccuracy	..	..	..	..	..	20	..	..	..	20	60
20. Coupling inaccuracy or damage	10	20	10	20	30	10	..	..	..	..	..
21. Rotor & bearing system critical	..	..	..	100	..	..	..	..	..	..	..
22. Coupling critical	..	..	..	100	Also make sure tooth fit is *tight!*						
23. Overhang critical	..	..	..	100	..	..	..	..	..	..	..
Structural resonance of: { 24. Casing	..	10	..	70	10	..	10	..	..	..	..
25. Supports	..	10	..	70	10	..	10	..	..	..	..
26. Foundation	..	20	..	60	10	..	10	..	..	..	..
27. Pressure pulsations	Most troublesome if combined with resonance									100	..
28. Electrically excited vibration	..	..	..	..	..		..	..	..	..	..
29. Vibration transmission	..	..	..	..	..		..	..	..	90	..
30. Valve vibration	..	..	..	..	..		..	..	..	..	100
Problem	The section below is meant to identify basic mechanisms										
31. Sub-harmonic resonance	Rare — Look for aerodynamic origin (seals)						←————100————→			..	..
32. Harmonic resonance	..	..	..	←————100————→			..	..	..	..	..
33. Friction induced whirl	80	10	10	..	..	..	..	..	..	..	..
34. Critical speed	..	..	..	100	..	..	..	..	..	..	..
35. Resonant vibration	..	..	..	100	..	..	..	..	..	..	..
36. Oil whirl	..	100		Watch for aerodynamic rotor-lift (partial admission, etc.)							
37. Resonant whirl	..	100	..	..	..	..	..	..	..	..	..
38. Dry whirl	..	..	..	..	..	..	..	..	..	..	100
39. Clearance induced vibrations	10	80	10	..	..	..	..	..	..	..	..
40. Torsional resonance	..	..	..	40	20	20	..	..	..	20	..
41. Transient torsional	..	..	..	50	..	..	..	..	..	50	..

Numbers indicate percent of cases showing above symptoms, for causes listed in vertical column at left.

Table 5-1 continued

Vert.	Hor.	Axial	Shaft	Bearings	Casing	Foundation	Piping	Coupling	Stays same	Increases	Decreases	Peaks	Comes suddenly	Drops out suddenly	Stays same	Increases	Decreases	Comes suddenly	Drops out suddenly
40	50	10	90	10	..	..	..	..	..	100	..	Peaks at critical	..	..	..	..	100	..	..
↓	↓	↓	↓	↓	..	..	..	..	..	100	..		..	..	..	..	..	..	..
					..	..	..	..	30	60	5		5	..	30	5	50	5	10
					..	..	..	..	30	50	5		5	10	30	5	50	5	10
					..	..	..	..	40	60	..		..	..	40	..	60	..	..
			40	30	10	10	10	..	20	80	..		..	..	20	..	80	..	..
30	40	30	80	10	10	..	..	..	10	70	..		10	10	10	..	70	10	10
30	40	30	70	10	20	..	..	..	10	40	10		20	20	10	..	50	20	20
20	30	50	80	10	..	..	..	..	20	30	10		20	20	20	..	40	20	20
20	30	50	80	10	..	..	..	..	20	40	..		20	20	20	..	40	20	20
40	50	10	90	10	..	..	..	..	40	50	10		..	..	40	10	50	..	..
30	40	30	70	20	10	..	..	..	10	50	10	↓	20	10	10	10	50	10	20
40	50	10	50	20	20	20	..	..	..	10	..	..	90	..	..	..	10	..	90
40	50	10	40	30	30	..	..	..	..	40	..	50	10	..	..	..	40	..	10
20	30	50	60	20	20	..	..	..	20	50	10	..	10	10	20	10	50	10	10
40	50	10	60	20	20	..	..	..	..	..	..	..	90	10	..	..	..	10	90
40	50	10	80	10	10	..	..	..	..	..	..	..	90	10	..	..	..	10	90
40	50	10	70	20	10	..	..	..	..	..	..	..	90	10	..	..	..	10	90
40	50	10	50	20	30	..	..	..	..	..	..	..	90	10	..	..	..	10	90
30	50	20	80	10	10	..	..	..	20	20	20	20	10	10	20	20	20	10	10
30	40	30	70	20	..	..	..	10	10	20	..	20	Loose sleeve, friction or dirt 40 in teeth 10		10	..	20	10	40
40	50	10	70	30	..	..	..	..	..	20	..	80	..	..	..	..	20	..	..
20	40	40	10	10	..	..	..	80	..	20	..	80	..	If loose	..	..	20	50 If loose	..
40	50	10	70	10	..	..	..	20	..	30	..	70	..	..	..	..	30	..	..
40	50	10	..	40	40	10	10	..	..	20	..	80	..	..	..	..	20	..	..
40	50	10	..	20	50	20	10	..	..	20	..	80	..	..	..	..	20	..	..
30	40	30	..	10	40	40	10	..	..	20	..	80	..	..	..	..	20	..	..
30	40	30	Can excite whirls or resonance		30	30	40	..	90	10% — Depending on origin of disturbance					90	10 ————→			
30	40	30	↓		40	40	20	..	90	..	..		..	..	90	..	..	..	..
30	40	30	↓	↓	40	40	20	..	90	..	..	↓	..	..	90	↓	..	..	..
30	40	30	..	..	80	10	10	..	80	..	..	..	10	10	80	..	..	10	10
30	30	40	20	80 If bearing is excited	20	20	20	..	..	20	..	20	30	30	..	..	20	30	30
40	40	20	20	10	10	30	30	..	20	20	..	60	..	..	20	..	20	..	..
40	50	10	80	20	..	..	..	..	..	..	..	..	90	10	..	..	..	10	90
40	50	10	60	40	..	..	..	..	..	20	..	80	..	..	..	..	20	..	..
40	40	20	20	10	20	30	20	..	..	20	..	80	..	..	..	..	20	..	..
40	50	10	80	20	..	..	..	..	..	..	..	..	100	..	..	..	..	..	100
40	50	10	20	20	20	20	20	..	..	..	..	..	80	20	..	..	..	20	80
30	40	30	40	20	20	10	..	10	..	..	..	..	80	20	80	..	..	..	20
40	50	10	70	10	10	..	..	10	..	..	..	..	80	20	20	..	..	20	60
Torsional	..	100	Lateral amplitude 40	40	..	..	..	10	..	20	..	30	30	20	20	..	..	20	30
..	↓	..	Torsion 100	40	40	..	..	10	..	..	..	50	30	20	..	..	..	30	20

Comments on Table 5-1
(Numbers correspond to "Cause of Vibration" in Table 5-1.)

1. Long, high-speed rotors often require field balancing at full speed to make adjustments for rotor deflection and final support conditions. Corrections can be made at balancing rings or at coupling bolts.

2. Bent rotors can sometimes be straightened by "hot-spot" procedure, but this should be regarded as a temporary solution because bow will come back in time and several rotor failures have resulted from this practice. If blades or discs have failed check for corrosion-fatigue, stress-corrosion, resonance, off-design operation.

3. Straighten bow slowly, running on turning gear or at low speed. If rubbing occurs, trip unit immediately and keep the rotor turning 90° using a shaft wrench every 5 minutes until the rub clears, then resume slow run. This may take 12 to 24 hours.

4. Often requires complete rework or new case but sometimes a mild distortion corrects itself with time (requires periodic internal and external realignment). Usually caused by excessive piping forces or thermal shock.

5. Usually caused by poor mat under the foundation or thermal stress (hot spots) or unequal shrinkage. May require extensive and costly repairs.

6. Slight rub may clear but trip the unit immediately if a high-speed rub gets worse. Turn by hand until clear.

7. Unless thrust bearing has failed this is caused by rapid changes of load and temperature. Machine should be opened and inspected.

8. Usually caused by excessive pipe strain and/or inadequate mounting and foundation. But is sometimes caused by local heat from pipes or the sun heating the base and foundation.

9. Most trouble is caused by poor pipe supports (should use spring hangers), improperly used expansion joints, and poor pipe line-up at casing connections Foundation settling can also cause severe strain.

10. Bearings may become distorted from heat. Make a hot check, if possible, observing contact.

11. Watch for brown discoloration which often precedes recurring failures. This indicates very high local oil film temperatures. Check rotor for vibration. Check bearing design and hot clearances. Check condition of oil, especially viscosity.

12. Check clearances and roundness of journal, as well as contact and tight bearing fit in the case. Watch out for vibration transmission from other sources and check the frequency. May require anti-whirl bearings or tilt-shoe bearings. Check especially for resonances at whirl frequency (or multiples) in foundation and piping.

13. This can excite resonances and criticals and combinations thereof at two times running frequency. Usually difficult to field balance because when horizontal vibration improves, vertical vibration gets worse and vice versa. It may be necessary to increase horizontal bearing support stiffness (or mass) if the problem is severe.

14. Usually the result of slugging the machine with fluid, solids built up on rotor, or off-design operation (especially surging).

15. The frequency at rotor support critical is characteristic. Discs and sleeves may have lost their interference fit by rapid temperature changes. Parts usually are not loose at standstill.

16. Often confused with oil whirl because the characteristics are essentially the same. Before suspecting any whirl, make sure everything in the bearing assembly is absolutely tight with an interference fit.

17. This should always be checked.

18. Usually involves sliding pedestals and casing feet. Check for friction, proper clearance and piping strains.

19. To obtain frequencies, tape a microphone to the gear case and record noise on magnetic tape.

20. Loose coupling sleeves are notorious troublemakers, especially in conjunction with long, heavy spacers. Check tooth fit by placing indicators on top then lift by hand or a jack and note looseness (should not be more than 1-2 mils at standstill, at most). Use hollow coupling spacers. Make sure coupling hubs have at least 1 mil/inch interference fit on shaft. Loose hubs have caused many shaft failures and serious vibration problems.

21. Try field balancing; more viscous oil (colder); larger, longer bearings with minimum clearance and tight fit; stiffen bearing supports and other structures between bearing and ground. This is basically a design problem. It may require additional stabilizing bearings or a solid coupling. It is difficult to correct in the field. With high-speed machines, adding mass at the bearing case helps considerably.

22. These are criticals of the spacer-teeth-overhang subsystem. Often encountered with long spacers. Make sure of tight-fitting teeth with a slight interference at standstill and make the spacer as light and stiff as possible (tubular). Consider using a solid or membrane coupling if the problem is severe. Check coupling balance.

23. Overhang criticals can be exceedingly troublesome. Long overhangs shift the nodal point of the rotor deflection line (free-free mode) towards the bearing, robbing the bearing of its damping capability. This can make critical speeds so rough that it is impossible to pass through them. Shorten the overhang or put in an outboard bearing for stabilization.

24. Casing resonance is also called "case-drumming." It can be very persistent but is sometimes harmless. The danger is that parts may come loose and fall into the machine. Also, rotor/casing interaction may be involved. Diaphragm drumming is serious, since it can cause catastrophic failure of the diaphragm.

25. Local drumming is usually harmless but major resonances, resulting in vibration of the entire case as a unit, are potentially dangerous because of possible rubs and component failures, as well as possible excitation of other vibrations.

26. Similar problems as in 24 and 25 with the added complications of settling, cracking, warping and misalignment. This cause may also produce piping troubles and possible case warpage. Foundation resonance is serious and greatly reduces unit reliability.

27. Pressure pulsations can excite other vibrations with possible serious consequences. Eliminate such vibrations using restraints, flexible pipe supports, sway braces, shock absorbers, etc., plus isolation of the foundation from piping, the building, basement and operating floor.

28. Occurs mostly at two times line frequency (7,200 cpm), coming from motor and generator fields. Turn the fields off to verify the source. Usually harmless, but if the foundation or other components (rotor critical or torsional) are resonant, the vibrations may be severe. There is a risk of catastrophic failure if there is a short circuit or other upsets.

29. This can excite serious vibrations or cause bearing failures. Isolate the piping and foundation and use shock absorbers and sway braces.

30. Valve vibration is rare but sometimes very violent. Such vibrations are aerodynamically excited. Change the valve shape to reduce turbulence and increase rigidity in the valve gear. Make sure the valve cannot spin.

31. The vibration is exactly one-half, one-quarter, one-eighth of the exciting frequency. It can only be excited in nonlinear systems, therefore, look for such things as looseness and aerodynamic or hydrodynamic excitations. It may involve rotor "shuttling." If so, check the seal system, thrust clearances, couplings, and rotor-stator clearance effects.

32. The vibrations are at 2x, 3x, 4x exciting frequency. The treatment is the same as for direct resonance: change the frequency and add damping.

33. If the cause is intermittent, look into temperature variations. Usually the rotor must be rebuilt, but first try to increase stator damping, add larger bearings (tilting-shoe), increase stator mass and stiffness, improve the foundation. This problem is usually caused by maloperation such as quick temperature changes and fluid slugging. Use membrane-type coupling.

34. This is basically a design problem but is often aggravated by poor balancing and a poor foundation. Try to field-balance the rotor at operating speed, lower oil temperature, and use larger and tighter bearings.

35. Add mass or change stiffness to shift the resonant frequency. Add damping. Reduce excitation, improve system isolation. Reducing mass or stiffness can leave the amplitude the same even if resonant frequency shifts, because of stronger amplification. Check "mobility."

36. Stiffen the foundation or bearing structure. Add mass at the bearing, increase critical speed, or use tilting-shoe bearings (which is the best solution). First, check for loose fit of bearings in bearing case.

37. Same comments as 36 with additional resonance of rotor, stator, foundation, piping, or external excitation, find the resonant members and the sources of excitation. Tilting-shoe bearings are the best. Check for loose bearings.

38. Sometimes you can hear the "squeal" of a bearing or seal. But frequency is usually ultrasonic. Very destructive. Check for rotor vanes hitting the stator, especially if clearances are smaller than the oil film thickness plus rotor deflection while passing through the critical speed.

39. Usually accompanied by rocking motions and beating within clearances. It is serious especially in the bearing assembly. Frequencies are often below running frequency. Make sure everything is absolutely tight, with some interference. Line-in-line fits are usually not sufficient to positively prevent this type of problem.

40. This problem is very destructive and difficult to find. The symptoms are: gear noise, wear on the back side of teeth, strong electrical noise or vibration, loose coupling bolts, fretting corrosion under the coupling hubs. There is wear on both sides of coupling teeth, and possibly torsional-fatigue cracks in keyway ends. The best solution is to install properly tuned torsional vibration dampers.

41. Similar to Item No. 40 but encountered only during startup and shutdown because of very strong torsional pulsations. It occurs in reciprocating machinery and synchronous motors. Check for torsional cracks.

The signal-to-noise ratio on a standard AM tape recorder is not bad. It is quite good for most data at −30 dB. The great drawback, of course, is the roll-off frequency which occurs at 50 to 60 Hertz on most AM tape recorders. Roll-off frequency is the condition of low frequency signal which becomes attenuated by the device that is taking the data and playing it back. On AM tape recorders you will find that on playback at a machine speed less than 50 cycles, which is equivalent to 3,000 rpm, the data will be displayed at a lesser value than it was actually recorded, thus giving you an error unless you know the amount of attenuation so that you might apply a correction. Fortunately, this fall-off is gradual, then it drops quickly (see Fig. 5-5).

You will obtain better fidelity or a higher frequency response range of data if a faster recording speed is used. For instance, 7½ inches per second will allow you to record frequency ranges higher than if 3¾ inches per second were used. For practical purposes, we generally do not use the 1⅞-inch per second speed on an AM tape recorder for taking data in the field no matter how slow the unit is running; however, if a long record time is very important and the speed is slow, for instance, down in the 3,000-6,000 rpm range, then satisfactory data can be obtained at 1⅞ inches per second.

The AM tape recorder is inexpensive compared to an FM recorder. Most AM recorders will cost about $150 to $200 per channel, whereas a good first-quality FM tape recorder will cost about $1,000 per channel. Try not to equate these two together because the FM recorder, being more expensive, is also capable of recording frequency levels down to zero speed (specifications will say down to DC levels). My experience in tape recorders has been with two- and four-channel units built by Sony to two- and four-channel FM and direct units built by the Dallas Instrument Company in Dallas, Texas, and a 14-channel FM or direct tape recorder built by Minneapolis Honeywell with the preamp and conditioning panels and the patchboard network built by the Bently Nevada Corporation Instrument Section on contract. To give you an idea on costs for these three ventures, (see Fig. 5-4) the four-channel Sony tape recorder (modified) cost about $650 at the time of purchase. The four-channel FM direct, battery-operated, Dallas Instrument recorder, which is used almost daily, will now cost $3,600. The Minneapolis Honeywell recorder with proper conditioning, amplifiers, and playback facilities cost us $20,000 in 1972. It is always dangerous to give pricing because pricing escalates almost semiannually.

To calibrate an AM tape recorder, use an oscillator to apply a signal to the channel-one tape recorder input with an oscilloscope displaying that signal. AM tape output channel (or playback channel) could be recorded on a second channel of the oscilloscope. The gain or sensitivity pot can be adjusted so that the input, for example, one volt peak-to-peak at 60 cycles per second, equals one volt peak-to-peak at 60 cycles per second on playback. Once this has been adjusted to the proper setting and locked, then the tape should be rewound and the signal played back. Again, see that the playback mode gives the same recorded signal, e.g., one volt peak-to-peak at 60 cycles

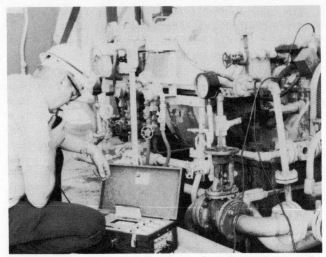

Fig. 5-4—A battery powered FM-direct cassette tape recorder improves the speed and simplicity of multiple data taking with voice narration.

per second. If the playback does not reach the input value, increase the gain adjustment. If it plays back too strongly, reduce gain adjustment, and through a quick trial and error procedure, the setting will be located in short order. Other channels may be calibrated in the same manner. This final setting should be locked, scribed and taped (match-marked with a pencil on the knob and the faceplates) so that if someone inadvertently moves the gain knobs they can be returned. (See Fig. 5-5.)

Since the AM tape recorder is so easy to use, the bias voltage that exists on a proximity system can be used without any problem. However, if you switch to an FM tape recorder, the bias voltage will saturate the tape without any vibration signals being recorded and all the data will be lost. To prevent this from happening, insert a capacitor in series with the input signal. Our previous articles suggest the limit range of that capacitor and that the capacitor will block the DC voltage and impress on the tape recorder only changing voltage, which will be the

AC component. It will be recorded at its true value, and only when that value reaches the saturation of the FM tape recorder will the tape head be saturated.

If you have an FM tape recorder, you will need a multiple range of attenuations of the signal, should it be too strong, to keep from saturating the tape; or, conversely, if the signal is too weak, an amplifier is needed to boost the signal in one range or another. There are five voltage ranges for the recorder shown in Fig. 5-6. The recorder can be used in an AC mode or a DC mode. DC mode implies that the signal is coming straight into the recorder without a capacitor in series. This particular recorder is quite useful in that it uses standard cassette film, which is not 1/4 inch but, rather, about 0.140 inch wide or slightly greater than 1/8 inch. Four channels of data can be recorded on these cassette packages. Further, it is a unit that operates on battery and does not need AC power. It has a frequency response range of 1 khz in the FM mode (60,000 rpm for a machine) or in the direct mode (signal is *not* modulated and then demodulated later from a carrier frequency), the frequency response range is 10 khz (600,000 rpm).

Data can be recorded on an FM recorder using a summing amplifier to balance the bias voltage, e.g. −7 volts with a +7 volts giving a 0 volt record at zero speed. Using this DC coupling arrangement, two probes can record shaft vibration while simultaneously recording shaft position (attitude) in the bearings (see Fig. 7-18).

Most FM tape recorders on the market have no preamplifiers. These must be purchased as an accessory. In recent years, several manufacturers, such as the Tanzberg, Hewlett-Packard, Lockheed, and others, have incorporated the preamplifier packages and alternate coupling features as a part of the tape recorder which is surely a step in the right direction. It would be good if the electronic builders would request from the user some of the features that are necessary so that the standard instrument package could be provided without a series of special features being required on every purchase that is placed.

Fig. 5-5—Calibration of an AM tape recorder is simple. A signal is applied by an oscillator shown on the far left. The record and playback signal is observed on the scope shown in the center. Sensitivity (per channel) is adjusted on the recorder with a screw driver.

Fig. 5-6—A battery driven FM-direct tape recorder has 5 voltage ranges with 3 KHz (FM) or KHz (direct) frequency response range with voice interrupt.

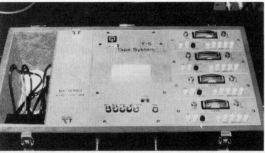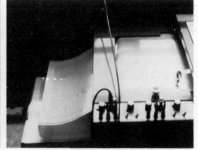

Fig. 5-7—Data can be field recorded on tape (center) processed through a Real Time Analyzer (left) (Fast Fourier Series) and recorded on an X-Y or X-Y-Y' (phase) recorder (right) for quick and distinct spectographs.

Calibration of an FM recorder must follow the particular instructions of the manufacturer, because the circuitry is much more complex in an FM tape recorder than an AM recorder. However, you will generally find, with a little practice, that your own calibration procedures are much easier to follow than those prescribed by the manufacturer.

If you take phase data, and surely you do, phase distortion can be a very serious matter on tape recorders. I have found that if there is a single record head and one channel is being used to record a phase mark or one event pulse per turn of the rotor shaft, phase distortion in one revolution from that signal to any one of the other recorded signals is not greatly affected except for some slight skew of the tape across the head or other defects—but this is not a serious matter if one record head is being used. However, consider the 14-channel tape recorder which has seven channels on one record head and the other seven channels on a second record head. If a channel on one record head is used for phase data then channels that are recorded on the same head will not have much phase distortion in it, but any channel that recorded on the other record head may suffer anywhere from, typically, 90° to 170° phase distortion on playback of the data recorded. This is a very serious error to a mechanical engineer wanting to know phase angle information. On our 14-channel recorder, the odd channels are on one record head and the even channels are on another record head. One of the improvements that we can make is to put our phase reference data on an odd and even channel that occurs in the center of the record head. While this reduces our error, it makes us waste an extra channel to get phase data. So my 14-channel tape recorder ends up being a 12-channel data recorder with phase.

The limitation imposed on the recorder builder is that he can only put so many channels on one record head, otherwise he will suffer the effects of crosstalk (one channel's data will be recorded on an adjacent channel; or, the signal-to-noise ratio will be reduced). A preferred signal-to-noise ratio for tape recorders or any other instrument is −60 dB; however, we feel quite happy when we obtain something in the order of −35 to −40 dB signal-to-noise ratio.

I reemphasize the importance of a tape recorder in vibration analysis, because it can record data on many channels while a person is able to view only so much data at one time. Oftentimes you are looking at one channel when something very bad is occurring on another channel. Perhaps the most important feature of a tape recorder is its use in analysis review following a machine failure. The tape can be replayed and the whole story can be reviewed in great detail. The tape speed can be slowed down, slow motion if you will, and the data carefully analyzed when many things may have happened in a very short period of time. Remember, however, that in slowing down the tape speed, the frequency information is changed in the same ratio. It is often easy to assume a condition occurred at one frequency when it really was at twice that frequency because the recorder speed has been slowed down.

In our plant's normal preventive maintenance program, the information is recorded both from proximity probes and from seismic pickups placed on the machinery. Three-case machine data can be recorded in about 18 minutes direct to a battery-operated tape recorder without requiring power cables and other appendages which become a handicap in the essence of time. Recorded data are then brought back to the lab and played back through a real-time analyzer. Each channel is then recorded into the individual analyzer. Each channel is then recorded into the individual discrete frequencies on an X-Y-Y' recorder. (See Fig. 5-7.) Frequency amplitude analysis is used for study or for recording in the equipment history book indicating the trend in level and frequency from one period to another. This is a great benefit, as it unscrambles a lot of detail into the important peaks and valleys so one can observe and analyze very quickly. For example, in our plant it is no problem to analyze ten machines in one day using this procedure; whereas, the slow, precarious manual tuning of a narrow band vibration analyzer would limit us to three machines in one working shift. In addition, data evaluation would take at least an equivalent period of that time because the high values and the significant frequencies must be selected by many minutes or hours of study from all the data.

I hope that a tape recorder will be on the market for nine channels of data on ¼ inch tape, battery-operated with AC and DC coupling, preamplifiers, a signal-to-noise ratio of −40 dB or better (with a level meter indicating each of the channels being used), available as an AC or DC coupled unit, voice interrupt, equipped with standard

7-inch reels with record speeds of 3¾ inches, 7½ inches, 15 inches and possibly 30 inches per second, while maintaining a price package of no more than $1,000 per channel. Specifying nine channels would allow phase data to be applied to one center channel, which would not be a data channel, and the edge track (or data interrupt voice channel) would be used for voice. Two points at two bearings or the equivalent of four bearings total data could be taken. For example, on a two-case machine with two bearings, or conversely a four-case machine including a gearbox, we could take data at one point per bearing and one channel would be available to take data alternately from the paired sensor point at each bearing during the data-taking cycle. At present, this is not possible because record heads are either in the two, four, seven, eight, or higher channels per record head. The best compromise at present for a narrow tape would be eight channels. The current seven channels on the market really reduce the data channels to six for all practical purposes, and that becomes a little difficult.

One more caution might be brought out in purchasing FM recorders. IRIG (Intermediate Range Instrument Group) units should be purchased which follow these specifications so that if a plant located in one part of the country wanted to receive tapes from other plants in their company but, at other locations, the use of data tapes recorded on other tape recorders would only be practical if the tape recorders met some basic standards for compatibility and playback. I would not wish to close this chapter without stating that tape recorders can be quite heavy and after you carry a unit around that weighs 100 pounds for any period of time, that 100 pounds soon feels like 200 pounds. To me, the weight restriction of best benefit would be for the tape recorder and its padded transit case for shipping should total no more than 74 pounds so that it would be easy to carry and could be shipped on the airlines as baggage rather than air freight.

LITERATURE CITED

[1] Sohre, J.S., "Operating Problems with High-Speed Turbomachinery Causes and Corrections," ASME Petroleum Mechanical Engineering Conference, Dallas, Sept. 1968.

Critical Speeds
and Mode Shapes

All turboequipment operates at certain speeds where the forcing or self-exciting frequencies are sympathetic (at synchronous or harmonics) to the resonances of a rotor, bearing, support system. When this resonance occurs at a finite operating speed, that speed is referred to as a critical speed.

Some simple examples may be helpful in understanding the many variables that occur in modern turbomachines of today. Speeds are now in the $N \times 10^4$ rpm range. Rotors weigh several tons. Bearing spans may be 115-190 inches. Bearings are 4-8 inches and of many designs.

Translational mode. Consider a single mass typical of a single stage steam turbine, but on hard (high spring constant—$N \times 10^7$ lb./in.) bearings and short overhangs with consistent shaft diameters. The first critical will be a rigid mode (no reverse bending or flexures). This mode is often referred to as a translational, balance or bouncing mode. The modal analysis of such a rotor system, relative to centers of geometric beams, would look like Fig. 6-1.

The point where the rotor flexure crosses the geometric bearing centers is referred to as a nodal (node) point. As the spring constant at the bearing becomes softer (less lbs./in. or less clamping) the mode shape assumes less deflection or a pure translational mode (see Fig. 6-2).

Damping. Most bearing designs that become softer also provide more damping, e.g., a ball bearing with very hard stiffness, or spring rate and near zero damping, versus an elliptical journal (babbitted) bearing. The softer bearing (lower k value) decreases the critical speed (refer to Chapter 1), and the oil film thickness of a journal bearing, $l/d = \frac{1}{2}$-$\frac{3}{4}$, provides a "squish" in the oil film or a velocity restraining force providing damping of the rotor eccentricity, i.e., vibration or displacement value.

One could think of damping as radial mounted shock absorbers (buffer cylinders) or dash pots, (see Fig. 6-3).

For this reason, rotor dynamics illustrate the stiffness of a bearing with a symbol representing a spring

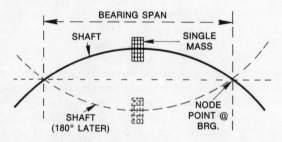

Fig. 6-1—First mode shape of example rotor system.

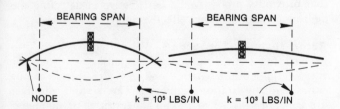

Fig. 6-2—A softer bearing produces less deflection.

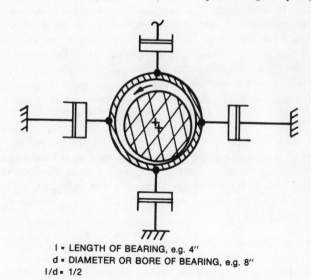

l = LENGTH OF BEARING, e.g. 4″
d = DIAMETER OR BORE OF BEARING, e.g. 8″
l/d = 1/2

Fig. 6-3—The oil film in bearings acts like shock absorbers or dash pots to dampen rotor eccentricity.

Fig. 6-4—Spring and dash pot symbols.

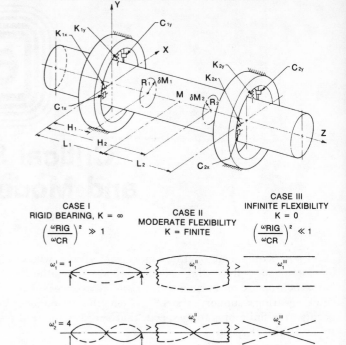

Fig. 6-5—Effect of bearing flexibility on critical speeds (Courtesy Dr. E.J. Gunter, Jr., University of Virginia).

(see Fig. 6-4A). Damping is represented by a dash pot symbol (see Fig. 6-4B).

Asymmetric bearings. Most computer programs provide the capabilities of four spring constants and four damping values. The reason for this is that the stiffness and damping values vary, e.g. the vertical stiffness, K_{yy}, of many bearings is greater than the horizontal stiffness, K_{xx}. If these values vary to a large degree, the bearing is said to be asymmetric and two criticals for each mode will be noted though often close together. This is confusing to a vibration analyst as it appears that the 1st and 2nd criticals were passed on the run when it was simply the horizontal first then the vertical first critical recorded. See Figs. 6-5 and 6-6 illustrating the effects of bearing flexibility on critical speeds.

Mode shape change. Returning to critical mode shapes again, as this first critical is passed the mode shape changes from translational or bouncing to that referred to as pivotal, conical or rocking modes. Multiple mode terms are cited, not to confuse the reader but to show different terms given to the same response used by different people as it appears that many rotor dynamicists each want to adopt their own terminology.

When this first critical is passed, the rotor mass (which was moving about the line of bearing centers) now reverts to the line of mass centers. If the balance was good, the material homogeneous in metal densities and all other parts (blading, shrouds, keys, etc.) geometrically even, this critical may not even be detected without unbalancing the rotor to $\simeq 1/10$-$1/5$g. (A 1,000-lb. rotor operating at 10,000 rpm with 16 gm.-in. of residual unbalance would create an unbalance force of 100 lbs. (1/10 of 1,000 lbs.) or 1/10 g.).

Because the rotor is now turning about the center of mass (center of gravity is on the center line of motion) the modal shape will change (see Fig. 6-7 on page 34). An increase in bearing damping will be effective if this motion occurs at the bearing (anti-node). A node (see Fig. 6-1) would not be affected by an increased damping change.

Phase change. When measuring phase, one will note a change in phase tending to be 180 degrees (if pure), see Fig. 6-8 on page 34. Before passing the first critical speed, the C.G. was turning about the bearing centers (refer to Chapter 4).

Reversal of the C.G. from the outer orbit to the center of rotation is cause for the 180-degree phase change. Fig. 6-9 on page 34 is a Bode plot of rotor response.

If all systems are pure, e.g. any mechanical bow is in phase with the residual unbalance with bearings symmetrical, the Fig. 6-9 plot is to be expected.

The next plot (Fig. 6-10) was taken on a five stage turbine with a thermal bow. The bow balanced out at 2,300 rpm indicating the bow was approximately 75 percent of the residual unbalance.

Pivotal mode. Turbomachinery designers generally have little fear of the pivotal (or conical) mode as the center of the machine has near zero deflections. Note the mode shape shown on the analysis of three critical speeds of a centrifugal compressor (see Fig. 6-11 on page 35). The node (zero deflection) of the rotor is at the center span. Large deflections occur at the bearing indicating that proximity probes would be effective in detecting the deflections and damping effective in reducing them.

Rotor response programs. When rotors are programed for critical speed and response programs, the rotor is laid out from one end, e.g. drive end, to the other end by station numbers. The station numbers are generally placed at: (1) changes in shaft diameter; (2) points of external load to the shaft; (3) centers of

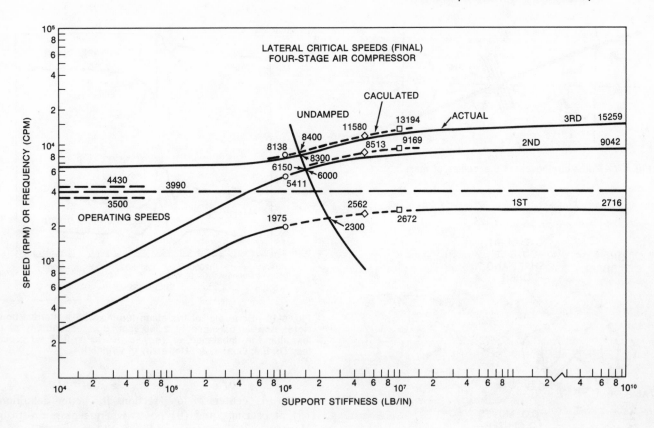

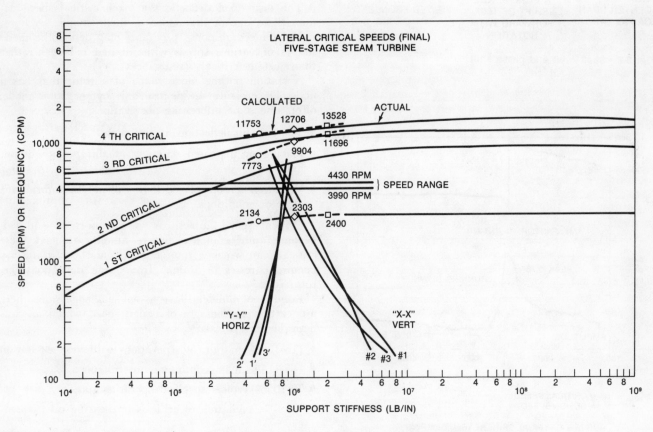

Fig. 6-6—Typical stiffness-critical map. This map is typical of *total* support stiffness effects on frequency.

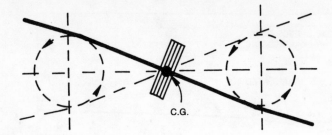

Fig. 6-7—Rotor turning about the center of mass.

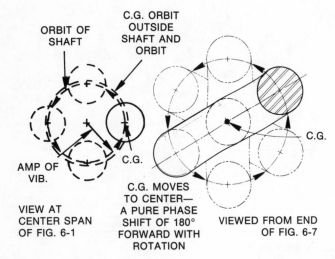

Fig. 6-8—Shaft orbit and phase shift.

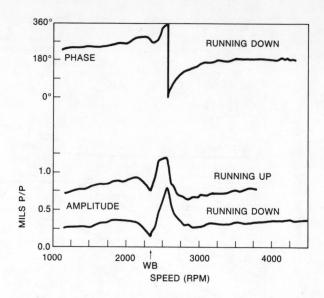

Fig. 6-10—Bode plot of five-stage turbine with a thermal bow. Note: Bow is balanced at 2,300 rpm. Bow eccentricity e_b is less than the unbalance, e_u, $[e_b = e_u \ (\omega_b/\omega_c^2 = 0.78 \ e_u]$ (Courtesy Dr. E.J. Gunter, Jr., University of Virginia).

mass; (4) centers of long sections for better definition; (5) at bearings, and (6) points where sensors are to be placed along the rotor (see Fig. 6-12 on page 36).

The display of deflections at each critical speed defines the shape of that mode or mode shape.

Unbalanced response programs may define the mode shapes of various speeds with inputed balances rather than just at criticals (resonances).

A station on the mode shape of a rotor that has a node, i.e. crossing the neutral axis by the rotor mode, offers one some interesting observations.

- There is no deflection at this point.

- Placing or increased damping at this point will not help.

- Vibration sensors placed here will not be informative of rotor conditions.

Critical speed analysis is very important and can become quite complex. Severe amplification at these criticals can wreck a rotor either from shear or reverse bending stresses or fatigue from large deflections, i.e. rotor rubs.

Equipment builders have a real challenge since there are certain percentages of design speed which are undesirable for criticals:

- 115-126% margin of separation to prevent excitation

- 105-115% max. continuous to trip speed

- 70-105% typical specified operating speeds

- 40-50% excitation of criticals by bearing oil frequencies

- -¼, ⅓, ½ speeds—Mathieu generators of sub-harmonic resonance.

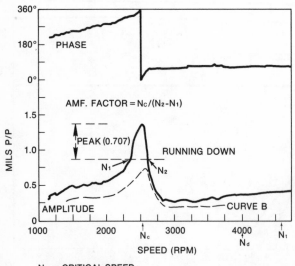

N_c = CRITICAL SPEED
N_d = DESIGN SPEED
N_t = TRIP SPEED
CURVE B IS TYPICAL OF INCREASED DAMPING

Fig. 6-9—Bode plot of rotor response—five-stage turbine.

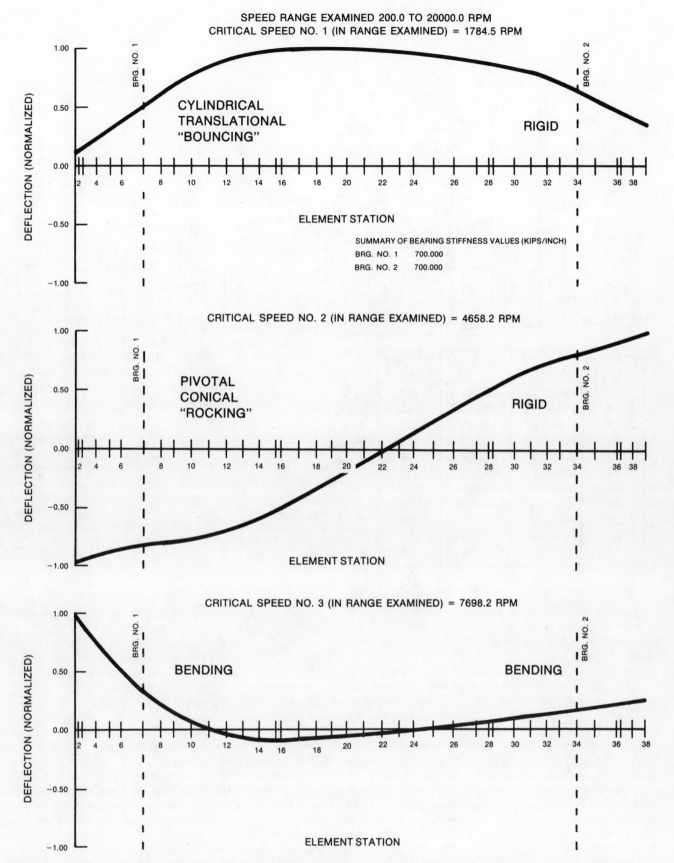

Fig. 6-11—Three critical speed plots of mode shapes on a centrifugal compressor. Note: a deflection of 1.00 merely represents maximum (100 percent) deflection on a non-dimensional scale. This is the mode shape *at* the critical only.

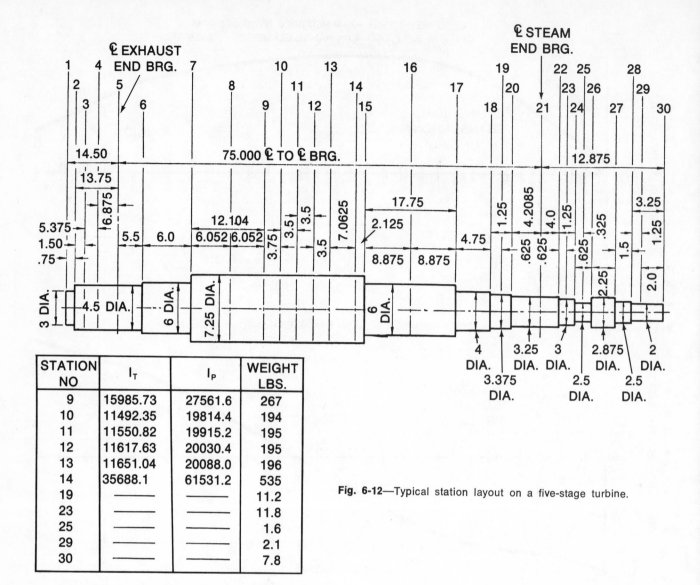

STATION NO	I_T	I_P	WEIGHT LBS.
9	15985.73	27561.6	267
10	11492.35	19814.4	194
11	11550.82	19915.2	195
12	11617.63	20030.4	195
13	11651.04	20088.0	196
14	35688.1	61531.2	535
19	———	———	11.2
23	———	———	11.8
25	———	———	1.6
29	———	———	2.1
30	———	———	7.8

Fig. 6-12—Typical station layout on a five-stage turbine.

For these reasons, many flexible rotor designers will often design first criticals to occur at 55-60% of design speed with second criticals over 120% of max. continuous speed.

Rotors with long spans and many stages become quite flexible and have caused several classic problems. Increasing bearing spring stiffness by stiffer designs or reducing bearing clearance is often the incorrect direction and only makes problems worse. Typical successes have been increased rotor clearances, larger shaft diameters, shorter spans or softer more damped (squeeze film) bearing designs.

An excitation of flexible rotor-balance criticals, while operating at design speed, can be more severe than when passing through the critical in coming to speed. For one thing, the damping effect of the drive torque has been greatly reduced.

Therefore, it seems reasonable to note the response of the criticals as a rotor passes each critical speed even though the critical is not in conflict with the specified operating speeds. A well damped (Low Q)

response is reassuring to the owner of a rotor system. A very narrow yet acceptable response of the critical causes concern that as the rotor loses its near perfect balance a sensitive rotor may be created. Some argue that a small unbalance is needed for damping (a velocity term).

Both points are valid. The bode plot during testing is very important and is now required in API Standards. It is unfortunate that aerodynamic loads, i.e. full load tests, cannot be performed on equipment before it leaves the manufacturer's facility.

Understanding the many things that can effect criticals is mind boggling. Many engineers, specializing in rotor dynamics, have spent the major part of their career studying this one area. There are so many variables—aerodynamic loading, gyroscopics, skewed elements, rotor assembly friction, asymmetric bearing, bowed rotors, friction rubs, seals acting as bearing, close clearance whirls, soft supports, soft foundations, soft soil condition, etc. Any or most any combination can greatly aggrevate the situation.

Instrumentation for Analysis

Instrumentation is necessary to perform vibration analysis. The amount and type of instrumentation depends on several factors:

• What does the analyst know from past experience?

• How complex is the analysis?

• How much data and how much time are required or allowed?

• What kind of data presentation is necessary?

• How much capital money is available for instrumentation?

• What are the climatic conditions where analytical data will be taken?

• How much transportation of instrumentation is involved?

• What repair facilities are available?

• How rugged should the instruments be built?

• What engineering terms are displayed and what calibration is needed?

This part will provide some guidelines to answer these questions. The author would not necessarily follow the same path in acquiring instrumentation today as in the past. Many instruments in use today were prototypes a short time ago. Vibration analysis instrumentation has developed greatly in the past 10 years. There seems to be a doubling of know-how and available commercial instruments every five years.

Instrumentation development was seriously pursued as rotating equipment became larger, operated at higher speeds, higher efficiency and without installed spare equipment. The demand for predictive maintenance and continuous protection monitoring is also higher.

API 670 Standard, "Noncontacting Vibration And Axial Position Monitoring System," for proximity sensors was published in June 1976. It is now being revised to add monitors using seismic-velocity and acceleration sensors.

Early vibration indicators. References to commercial instrumentation does not imply that many others could

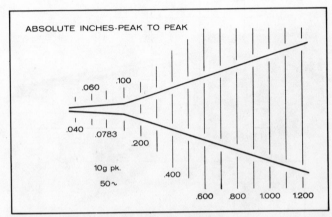

Fig. 7-1—Displacement calibration wedge.

not meet or exceed the performance discussed but the author has favorable experience with models illustrated or described.

Early indicators of vibration amounts were determined by visual indicators which varied, i.e., feel, sound, balanced objects such as coins, vibrating reeds and displacement calibration wedges (Fig. 7-1). There were visual indicators using the perception of sight and movement.

General Electric Co.'s Vibrometer came on the scene using a stylus and vibrating light beam (Fig. 7-2). This unit was followed by the hand-held Vibragraph by Askania and Davey (Fig. 7-2). The Davey Vibragraph used pressure sensitive tape traveling at 3 inches per second. Four attenuations, 1:1, 5:1, 10:1, 20:1 were provided, plus a time event marker at one-half second intervals. The tip can be either a Teflon button or Vee stick, often referred to as a "Fish Tail," made from Wood or Micarta. Drive power was obtained from a wound spring. This unit was very portable, data could be photocopied for reports, but required some knowledge of wave forms, e.g., to determine frequencies and frequency ratios. (Refer to Chapter 4 run traces taken from a Davey Vibragraph. Refer also to Chapter 5 for notes on taking field data.)

Available today are a number of small portable meters to read vibration levels. Most are battery powered. These meters can cost less than $500 or even less than $300. The Metrix 5160 meter (Fig. 7-3) is shown as a typical meter of this class. The more practical model reads any

Fig. 7-3—Metrix 5160 DV.

Fig. 7-2—General Electric Vibrometer (top) and Davey hand-held Vibrometer.

two of three parameters (A) acceleration in "g's"; (V) peak velocity, ipsp; or (D) displacement in peak-to-peak (mils). For the frequency range limit of 1 kHz or 60,000 rpm, the DV combination works well. This instrument is about the size of a two-cell flashlight, has several attachments and integrates an accelerometer output for circuitry.

Tunable analyzers. Investigating the market of tunable analyzers, one finds many features and price tags. Most analyzers can be obtained with battery or AC power but AC power is a necessary mode if a strobe light is provided or if the power regulation is taken from the battery, i.e., if the battery is dead, the analyzer will not function, even though AC power is available. Most newer models allow either a battery or AC power source and will operate on AC with a low battery on recharge.

There are a number of important features that should be considered when selecting this type of tunable analyzer:

1. Can it be used for field balancing?

2. Are the face markings permanent or will they wipe-off in service?

3. Is there a broad and narrow filter network?

4. How much does it weigh? Can it be taken as aircraft carry-on luggage?

Fig. 7-3—Metrix 5160 DV.

5. Will use in fog, light rain or heavy humidity be harmful to the circuits (i.e. is the case water tight)?

6. Are the circuits fungicide tested?

7. How many different sensors can be used? Are they provided?

8. Is the strobe light bright enough to see in open daylight?

9. Are the cables and connectors built strongly with strain relief against fatigue failures?

10. What is the frequency response range, i.e., the usable frequency range?

11. Are the vibration and frequency scales and units easy to read?

12. Is there a power voltage and frequency selector switch to allow use in United States, Europe, Asia, etc., e.g., 50 cps, 230V?

13. How stable are the circuits—warmup cycle, transcients, overloads?

14. Is there overload signal protection and indicators to prevent errors caused by amplifier overloads, i.e., false circuit harmonies at 1X, 2X, 3X . . . ?

15. Are there scope or recorder jacks?

16. Are accessories available?

17. Is it easy to calibrate?

18. Is it free from inadvertent (accidental) deflecting (bumping) of calibration controls?

19. Is it resistant to normal shocks from bumping, handling and shipping?

20. How much training is required for the analyzer operator? Can it be repaired easily? Where is the nearest service center? What is the normal repair time? Are loan units available during repairs?

Three small tunable analyzers with different advantages but filling most of the general acceptance criteria listed

Fig. 7-4—IRD 350.

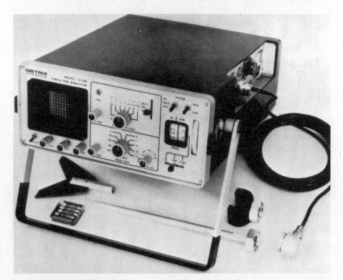

Fig. 7-5—Metrix 5115B.

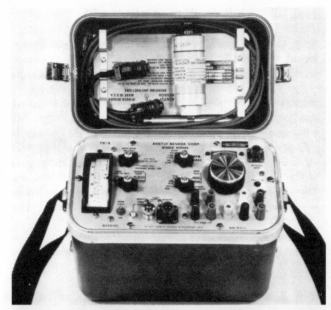

Fig. 7-6—Bently Nevada TK 8.

Fig. 7-7—IRD 360.

Fig. 7-8—Bently Nevada DVF 2.

Fig. 7-9—Spectral Dynamics 119B

above, are the IRD 350 (Fig. 7-4), Metrix 5115B (Fig. 7-5) and Bently Nevada TK-8 (Fig. 7-6).

Synchronous tracking filters came on the market next, to plot running speed resonance of a machine brought to operating speed through a certain critical or resonant speed. Notable were the IRD 360 (Fig. 7-7), Bently Nevada DVF 2 (Fig. 7-8). and Spectral Dynamics 119B (Fig. 7-9). Since these instruments plotted vibration amount, phase and rpm, they were also called trim balancers. Tracking rates would reach 10,000 rpm in 10 seconds or better. Later models integrated different sensors to read acceleration, velocity or displacement. Also, variable voltage calibration was included.

Since the Bode plot of speed, vibration and phase (Fig. 7-10) had become a significant part of machinery testing, the capability of tracking ½, 1 and 2 times running speed was included in the Bently DVF 2; which also provided notch, filter and filter-out modes. At least two manufacturers, Bently DVF 2 and IRD 360 Modified, had added features to allow polar (Nyquist) plots (Fig. 7-11) to be made.

Tracking rate is still one factor which must be evaluated if one is to record speed, amplitude and phase while

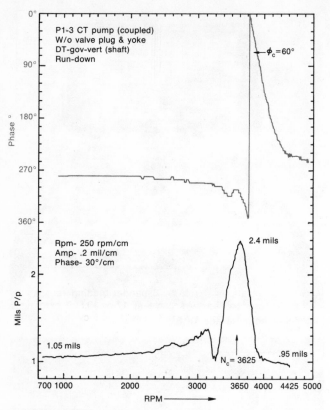

Fig. 7-10—Bode plot of steam turbine driven pump.

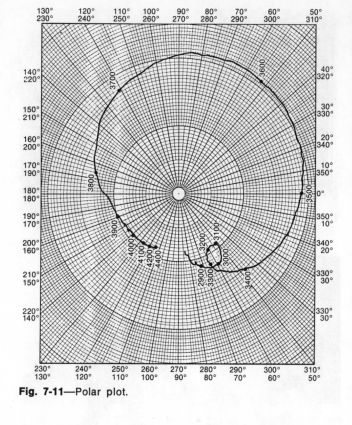

Fig. 7-11—Polar plot.

a machine is accelerating to speed or decelerating from speed. The Bently DVF 2 digital vector filter has the following tracking rates:

Speed Range rpm (Span)	Accel.	T (sec.)	Track Rate (rpm/sec.) (For 1% Error)*
5,000	Up	0.05	1,000
	Down	0.1	500
10,000	Up	0.05	2,000
	Down	0.10	1,000
20,000	Up	0.07	2,800
	Down	0.14	1,400
50,000	Up	0.16	3,000
	Down	0.32	1,500
100,000	Up	0.3	3,300
	Down	0.6	1,600

This unit uses two filter widths (120 rpm (2Hz) and 12 rpm (0.2Hz)) with respective Time Constants (TC) of approximately 0.2 sec. and approximately 2.0 sec. for amplitude and phase. The *slew* or track rate for X%

$$\text{error} = \frac{X\%/100(\text{Full Scale})}{\text{Time Constant (TC)}}$$

With the aid of AM, FM and digital tape recorders, a vibration analyst can do nearly anything necessary to determine operating conditions in turbomachines. With

* Note: Percent error is always applied to the full scale range. In speed, 1% error, at 10,000 rpm span, is 100 rpm.

minicomputer prices dropping and features being added, one should be able to perform a cascade (3D) display of all spectra from zero to operating speed with a minimum of effort and capital investment. The cascade display, shown in Fig. 7-15, has several synonymous displays, i.e., waterfall, raster, 3D and Campbell. The "Campbell" is improper, in my opinion, as the Campbell Diagram is a design display of operating speed and harmonics vs. known resonances—also referred to as an Interference Program. One is shown in Fig. 7-16.

Torsional analysis did not receive much attention in the past, but it is now coming into vogue. Because of energy costs, many turbomachines are being specified with synchronous motors and gear drives. Starting transient torques are exceeding steady load torques by 300-1000 percent.

Spectrum analyzers. Simultaneous with the tracking filter, the spectrum analyzer made revolutionary developments. Analyzers with the capability of memory, averaging, transient capture, peak hold, 20 kHz frequency response, -60 dB dynamic range, internal calibration, orbit plots, total engineering units, harmonic modes, photograph or plotting capabilities and many other features hit the market. The prices varied from $7,000-$20,000, without plotters.

Three builders with capable instruments in this line were the Spectral Dynamics 335/340 (Fig. 7-12), Bently's EMR (Fig. 7-13) and Nicolet's 444 (Fig. 7-14) (which has

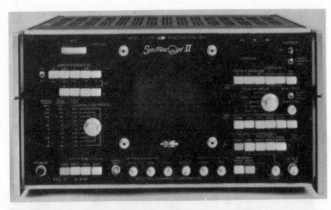

Fig. 7-12—Spectral Dynamics 335.

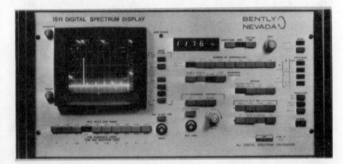

Fig. 7-13—Bently Nevada EMR.

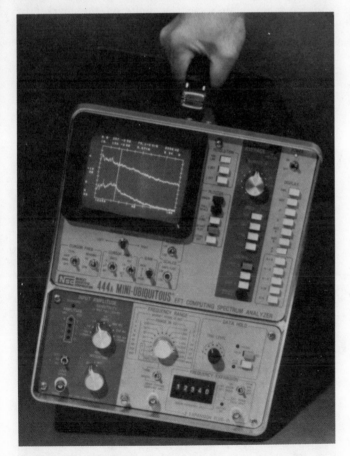

Fig. 7-14—Nicolet 444.

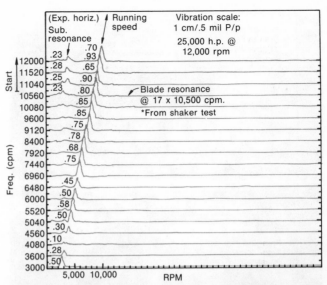

Fig. 7-15—Cascade display. Note: Expander blading resonance (by Shaker test) proved responsive at 17 x 10,500 cpm (i.e. rpm). There are 17 nozzles. Nozzle pass frequency is 17 x rpm.

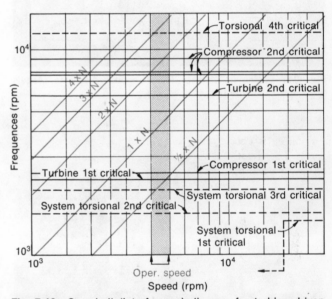

Fig. 7-16—Campbell (Interference) diagram for turbine driven compressor train. Note: Compressor 2nd critical intersects operating speed at twice running speed.

the most features). Spectral Dynamics could couple their Signature Ratio Adapter to the Real Time Analyzer to provide tracking and spectrum plus a wide range of harmonics of the running speed which is helpful on gear analysis. Nicolet could couple a digital tape recorder to its 444 model.

Most spectrum analyzers work on the basis of fast fourier transforms (FFT), or time compression principle, to increase data response speed. The FFT is gaining in popularity where two or more channels are incorporated or where phase data are not to be lost. Probably the most complete unit is offered by Hewlett Packard. However, the price tag may stagger available funds!

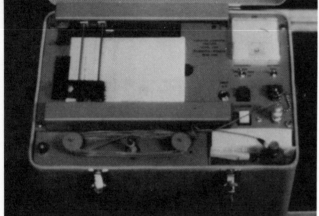

Fig. 7-17—Scientific Atlanta FM Torsional Analyzer Models 2429 and 2520 with plotter.

Fig. 7-18—Hewlett Packard Model 3968A eight channel, 14 in. tape FM/direct tape recorder.

Some manufacturers are incorporating torsional resonance features to existing concepts. It is unfortunate that Bell & Howell, CEC Division, no longer manufactures its original torsiograph sensor. Quan Tech, a subsidiary of Scientific Atlanta, manufactures an inexpensive and flexible FM unit (Fig. 7-17). The input sensor depends on tooth gap variations, with time, taken from a magnetic or eddy current probe looking at either gear teeth or an artificial stub gear extension to a rotor. The tooth gap variation, in velocity terms, is integrated for torsional deflections in degrees. There is a provision for the rotary sensor encoder but from a sensor made by others.

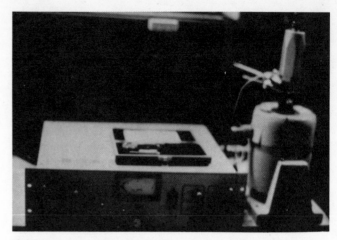

Fig. 7-19—Ling Shaker and amplifier with displacement sensors, velocity sensor, accelerometer and impedance head sensor under calibration check.

Tape recorders. One of the major problems in collecting data is the poor quality and lack of reliable FM/Direct tape recorders. It is bad enough that the cost of FM tape recorders varies from $1,000 to $2,000 per channel; but when you couple this to poor performance or severe frequency and phase distortion, it can easily aggravate the capabilities of the more reliable analyzers.

Builders of tape equipment varying from low cost to high cost are Dallas Instruments, Tanzberg, Hewlett-Packard, Lockheed, Honeywell, Ampex, Bell & Howell and Sangamo, with some new units on the market by B & K. Single data channels do not help. Nine data channels, including phase, are preferred (as described in Chapter 5). However, tape data heads are available with the following channels: 2, 4, 7, 8, 14, 28, . . . The eight channel Hewlett-Packard tape recorder (Bently Modified) shown in Fig. 7-18 uses ¼-in. tape; and with front panel calibration it has proven to be a helpful unit (priced under $10,000).

Other instrumentation. Good instrumentation depends on good calibration. Good calibration depends on proper wave or pulse generators, scopes, shakers, oscillators, DC signals, electronic counters, power supplies, etc.

The author has had good experience with a Ling Shaker and Amplifier at 55 lb-g* rating (Fig. 7-19), Hewlett-Packard Power Supplies, Wavetek Oscillators, Pioneer DC Voltage outputs (Quik-E), Bently Wobulators (Sine generator for proximity probes) and Counters, Monsanto Counters, Tektronic and Hewlett-Packard Scopes; plus B & K, and Endevco calibration and data accelerometers.

While the subject of practical plant instrumentation becomes quite broad and understandably commercial, it is hoped that the illustrations in this chapter primarily show the need for proper presentation of clear data. The manufacturers of such equipment are almost endless. Service facilities and performance, however, are key factors often overlooked by the purchaser.

* This means a 55-lb. weight can be excited to 1g. Conversely, a 5.5 lb. weight can be excited to 10g's. The heaviest velocity sensor I use is 1.5 lbs. and can be excited to 36g's; but an accelerometer sensor can be excited to more than 100 g's.

Vibration Severity

Vibration measurements of rotating equipment will always provide data, without exception. The real crux of the situation is whether the amount is acceptable, marginal or dangerous. It is supposed that these are the three major categories; however, many previous written guidelines break these categories into four or five levels. From a practical operating standpoint, four are sufficient: acceptable, marginal, schedule for shutdown repairs, shutdown immediate, but in an orderly fashion.

Then one presupposes that vibration severity is the test for selecting a category. For every two good guidelines, there is one class of equipment that becomes an exception. There are no absolutes. Newer equipment designs often require new guidelines.

The speed, rotor weight, rotor span, rotor stiffness, rotor components, type of bearings, type of seals and type of total support all contribute to the major category of severity.

It is for these reasons that you will see limitations placed on certain classes of equipment.

Severity charts. When T.C. Rathbone developed the original haystack charts, he was dealing primarily with turbines, operating at a medium speed with 2 or 3:1 ratio of shaft vibration to bearing housing (or pedestal) vibration. Other equipment, e.g., blowers, fans, pumps, were generally measured on the bearing with a seismic-velocity sensor and readout or a shaft-riding, hand-held Vibragraph.

IRD Mechanalysis (formerly International Research Development Corp.), along with Raydata (later to become Reliance) and the Philips (European) were early on the market with tunable analyzers. Fig. 8-1 severity chart was developed by Yates and reworked by IRD; however, it must be remembered that this chart was presented for bearing cap or pedestal reading. Blake's severity chart is also shown for comparison (Fig. 8-2).

Barrel type (vertically split) compressors were manufactured after the charts were developed. Here, the bearing is within the casing which could easily be 3-4-in. thick with an end plate of equal thickness housing the bearing. A barrel compressor is seen in Fig. 8-3.

Pipe line boosters represent a further deviation from the norm as the weight ratio of compressor-to-rotor reach 125:1 as seen in this 24-in. x 24-in. pipe line booster (also commonly used as a methanol or ammonia circulator compressor). See Fig. 8-4.

The data shown in Fig. 8-5 illustrate differences expected on data measured at the casing, in velocity (peak), vs. the shaft relative to bearing vibration taken simultaneously by vibration probes mounted at the same angular position (± 20 degrees). To make a fair comparison, the velocity in inches/second (peak) must be divided by the frequency with proper units., i.e.

$$\pi f / 1000 \text{ (where } f = \text{frequency, Hz)}$$

to give displacement in mils peak-to-peak. Note: V in./sec. peak/$[\pi f/1,000]$ = mils p/p.

Another severity chart (Fig. 8-6) for medium and high speed centrifugal compressors was published in 1970.

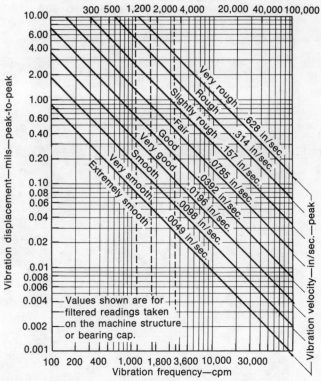

Fig. 8-1—General severity chart for bearing cap measurement.[1]

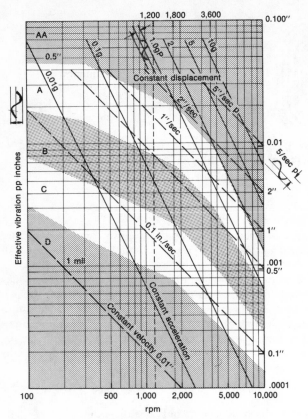

Fig. 8-2—Vibration chart for bearing cap measurement.[2]

Explanation of classes

AA Dangerous. Shut it down now to avoid danger.
A   Failure is near. Correct within two days to avoid breakdown.
B   Faulty. Correct it within 10 days to save maintenance dollars.
C   Minor faults. Correction wastes dollars.
D   No faults. Typical new equipment.

This is a guide to aid judgment, not to replace it. Use common sense. Use with care. Take account of all local circumstances. Consider: safety, labor costs, downtime costs.

Again, notice that the guidelines restricted this chart to equipment manufactured by Dresser Clark. This chart was the first the author had seen based on shaft vibration in mils peak-to-peak displacement measured relative to bearings. My experience parallels that of Dresser.

Two additional peak velocity curves have been superimposed over Fig. 8-6 to show the relative magnitude between the marginal vibration at 0.2 in./sec. peak on the IRD (Yates/Rathbone) chart vs. the Dresser chart of 1½ ips (peak) shaft relative motion.

Transmissibility is a ratio expressing the efficiency of a vibration motion through the bearing and case. If the transmissibility of steam turbines with separate bearing housings was 2:1 in the 1950s and if the velocity criteria of shaft readings on centrifugal compressors is up by 8, then it would seem reasonable that the barrel compressor transmissibility is only 2 x 8 = 16:1, which is about right.

Transmissibility, T = Shaft Vib./Housing Vib. = 16.

In structural isolation damping, one will often see an isolation ratio stated. This is simply the reciprocal of transmissibility, i.e., iso. ratio $1/T = 1/16$.

The object of isolation damping is to isolate or separate the vibration so it is not transmitted.

Measuring vibration with a sensor is affected by transmissibility. The significance of transmissibility becomes more complex in very high speed equipment, for example, turboexpanders, high speed pumps, etc. Do you measure on the case or rotor? It is my experience that it is best to measure internal rotor motion rather than case vibration in high speed service.

Fig. 8-7 shows an expander operating at 52,000-54,000 rpm in cold hydrogen service developing about 250 horsepower. The unit operates in a rigid pivotal mode (see Chapter 6) with each impeller overhung. The shaft re-

Fig. 8-3—Barrel type (vertically split) centrifugal compressor.

sembles your mother's kitchen rolling pin. The first mode is expected to be pivotal.

An ounce-inch of unbalance develops a force of 2½ tons. When Dr. Judson S. Swearingen, Rotoflow Corp., started designing this equipment, he also had to design and patent his own balancing equipment. To improve the sensitivity, he used a sensor which appears to be a pendulum mass operating in resonance at the balancing speed.

As stated in Chapters 4 and 5, there is no perfect sensor for all situations, and it is often necessary to use more than one type in critical situations. A separate part of this series has been devoted to the critical area of axial displacement on thrust movement monitoring. Using the same reasoning, no severity chart will fit all types of equipment. The vibration experienced in a crusher or mill would be too severe for a steam turbine. Other design provisions make allowances for this difference such as soft mounts, bearing damping and torsional tuning.

Gearing. I have always allowed an extra 0.1 ips (peak) velocity for gearing over other turbomachinery. Gearing generally contains short spans and has many exciting frequencies. Gear separating forces give additional high bearing loads at attitudes often not expected, e.g., horizontal plane loading. Transmissibility to the bearing is high in gear units. The API 613 (2nd Edition) Appendix gives some insite to the changes in critical speeds at no load vs. percentages of full load.

In gears, first and second order running speeds are expected (particularly with double helical gearing) because of asymmetry in pitch line runouts. Gear mismatching or misalignments can also cause severe axial motion in the pinion since the gear is normally retained by thrust bearings. Gear mesh frequencies, plus the first and second side bands, are important frequencies because deteriorations of the tooth mesh will show up at gear mesh frequencies. The gear mesh frequency is the number of teeth on a gear times its speed. The first side band of this frequency is one speed unit plus or minus from the gear mesh frequency. An increaser gear with a 2.5:1 gear ratio and input speed of 3,600 (60 Hz) rpm shown in Fig. 8-8 would have a gear mesh frequency of 144,000 cpm (40 x 3,600).

The first side bands would be 147,600 and 140,400 cpm (144,000 ± 3,600). The pinion gear side bands would be (144,000 ± 9,000). The gear's 2nd side bands would be 144,000 ± 2(3,600) or 151,200/136,800 cpm.

Gears are mechanical analogies to the electronics mixing or modulating circuits. Sums, differences and products of gear teeth frequencies can be prevalent, particularly the gear ratio divisor and multiplier of each shaft end speed. It also seems noteworthy that between the first and fifth orders of running speed and the gear mesh frequencies lie a band of intermediate frequencies which can be difficult, sometimes impossible, to define. Gear elements, the foundation and gear casing fundamental frequencies often lie in this band.

A concept of gear monitoring using accelerometers was described to me by John Mitchell, consultant (formerly with Endevco). The monitor has four band frequencies covering (1) low or gear speed ranges, (2) intermediate, (3) high gear mesh and (4) very high or acoustical emission range. Each frequency is selectable by a front panel five-position selector switch (the 5th switch is for the unfiltered signal).

My interpretation of vibration severity in most petrochemical equipment is shown in Table 8-1. Again, *no absolutes.*

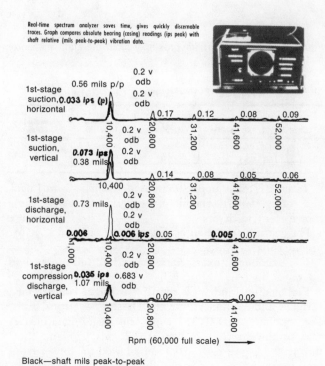

Fig. 8-5—Data on velocity (case) vs. displacement (shaft) reading on high speed centrifugal compressors.

Fig. 8-4—Circulator single stage 24 x 24 centrifugal compressor.

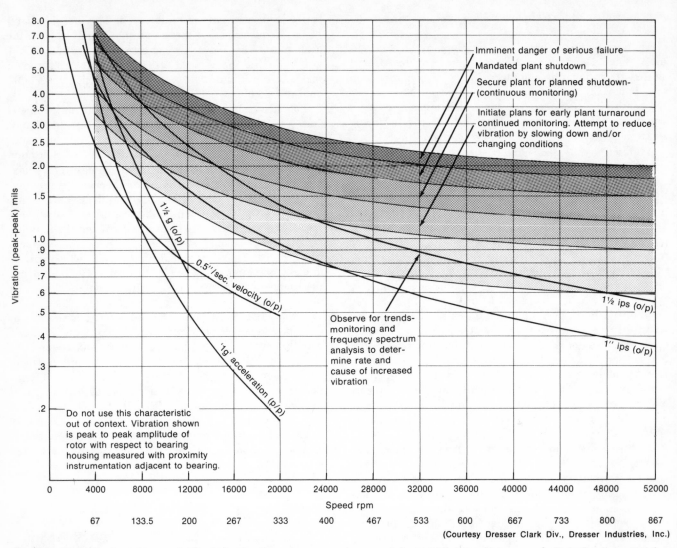

Imminent danger of serious failure
Mandated plant shutdown
Secure plant for planned shutdown (continuous monitoring)
Initiate plans for early plant turnaround continued monitoring. Attempt to reduce vibration by slowing down and/or changing conditions

1½ g (o/p)

0.5″/sec. velocity (o/p)

1½ ips (o/p)

1″ ips (o/p)

'1g' acceleration (p/p)

Observe for trends-monitoring and frequency spectrum analysis to determine rate and cause of increased vibration

Do not use this characteristic out of context. Vibration shown is peak to peak amplitude of rotor with respect to bearing housing measured with proximity instrumentation adjacent to bearing.

Vibration (peak-peak) mils

Speed rpm

(Courtesy Dresser Clark Div., Dresser Industries, Inc.)

Fig. 8-6—Severity chart for proximity shaft vibration measurement on certain centrifugal compressors.[3] (Velocity lines (o/p) added by C. Jackson)

TABLE 8-1—Bearing cap data

Smooth	Acceptable	Marginal	Planned Shutdown Repairs	Immediate Shutdown
0.1 ips (p) And Less	0.1 ips (p) To 0.2 ips (p)	0.2 ips (p) To 0.3 ips	0.3 ips (p) To 0.5 ips	0.5 ips (p)

Table 8-1 data apply to pumps, turbines, centrifugal and rotary compressors, motors, expanders, blowers, fans, centrifuges and dryers.

For gearing and reciprocating compressors, add 0.1 ips (p) to Table 8-1 values.

For bandbury mixers, hammer or ball mills, etc., add 0.1 to 0.2 ips (p).

For medium and high speed centrifugal compressors, turbines and expanders, use the Dresser Severity Charts. For example, on 10,000 rpm centrifugal compressors and turbines, set the alarm at 2.2 mils p/p and shutdowns at 4.2 mils p/p (+0.2 mil allowed for total runout tolerance).

Balancing severity. Good rotating equipment performance depends on good balance. This fact becomes sig-

nificant as speed and the number of components increase for a given rotor.

If a rotor operates below the first rigid balance critical, it is said to be a "stiff" or "rigid" shaft design. If it operates above this critical, it is referred to as a flexible rotor design even though it may be below the first "bending" mode (refer to Chapter 6 on Criticals).

Since the number of critical speeds affects the need for modal balancing, the number of critical speeds to be passed is important in obtaining the desired operating speed ranges. Many flexible shaft rotors operate above two or more criticals. It is easy to see then why N+2 balance planes are required by many engineers, i.e., number of criticals, N, plus two planes of correction.

It should be apparent looking at the mode shapes that an improper correction for the first mode may completely upset a rotor above the first critical.

If the rotor was balanced below the critical, i.e., typical of a balance machine speed, a different response might occur above the first critical because of a change in mode shape. For example, consider the rotor balanced as shown in Fig. 8-9.

This correction added in phase at disc 1 and disc 3 satisfies the first mode but may aggravate the second mode (see Fig. 8-10).

However, weights properly placed, say, in disc 2 for the first mode and coupled 180 degrees out of phase in discs 1 and 3 would not destroy the first mode and couple connect the second mode (see Fig. 8-11).

How much residual unbalance should be allowed and how much trial weight should be used when field balancing? In my experience, the answer to both questions is the same. The residual unbalance should not create an unbalance force exceeding 10 percent of the rotor weight at operating speed. If two balance planes are used (e. g., at each journal weight) journal weights or balance plane rotor weight can be substituted for the rotor weights. This is noted on most balance charts. A value in gram-inches, for example, is suggested with a note recommending ½ the value if two planes are used.

Unbalance forces can be thought of as $F_u = Ma$ or Unbalance Force = (Mass) (Acceleration).

Since Mass $= W/g$ and acceleration is $w^2 r$ then $F_u = (W/g) w^2 r$. This is often more properly expressed as Mew^2, i.e., (mass) (eccentricity) (w^2). Expressing this relationship in W (lbs.), g (386 in./sec.^{2}), w (rad./sec.) and r (in.) yields the Unbalance Force, F_u in lbs. This

value should be less than the rotor weight $(1g)$ and properly at 1/10 the rotor weight, $(1/10g)$.

The above expression could be stated in simpler units by rearranging the equation F_u (lbs.) = 1.77 (rpm/1,000)^2 oz.-in. or F_u (lbs.) = 1/16 (rpm/1,000)$_2$ gm-in.

Suppose a 1,000-pound rotor operates at 10,000 rpm, what should be the residual unbalance?

W = 1000 lbs.

$$F_u = (1/10)(1000) = 100 \text{ lbs.} = 1/16 \left(\frac{10,000}{1,000}\right)^2 \text{gm-in.}$$

$$\text{gm-in.} = \frac{100(16)}{(100)(1)} = 16 \text{ gm-in.}$$

What is the eccentricity (Fig. 8-12) from this unbalance?

$$\frac{16 \text{ gm.-in.}}{(1000)(454 \text{ gms/lb.})} = \frac{16 \text{ gm.-in.}}{454,000 \text{ gms}} = 35.24 \times 10^{-6} \text{ in.}$$

What sort of vibration is expected from this residual unbalance as measured with proximity probes?

Vibration, mils peak-to-peak (Fig. 8-13) = 2 × Amplitude in mils = 2 × Mass Center Displacement = (2) (eccentricity) = (2 × 35.24 × 10^{-6})/10^{-3} = 70.48 × 10^{-3} mils (0.07048 mils) or less than 1/10 mil.

Fig. 8-7—Turboexpanders.[4]

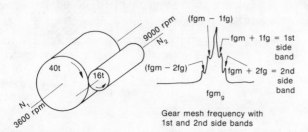
Fig. 8-8—Increaser gear and mesh frequencies.

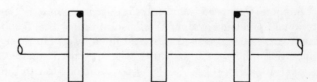
Fig. 8-9—Example of balanced rotor, first mode only.

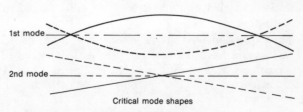
Fig. 8-10—Rotor's first and second mode shapes.

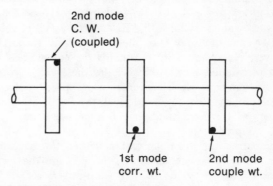
Fig. 8-11—Example of relocating weights.

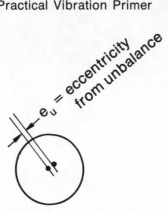

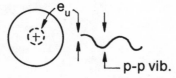

Fig. 8-12—Eccentricity from unbalance.

Fig. 8-13—Peak to peak vibration with residual unbalance.

This should be a satisfactory residual unbalance.

Can it be obtained on commercial balance machines? Yes. Most balancers guarantee a balance down to 25 microinches (25×10^{-6}). Balance for the example is for 35 microinches (35×10^{-6} inches).

Suppose the speed were only 5,000 rpm. What is the residual unbalance value? Well, substituting the 5,000 rpm where 10,000 is used will yield the answer. However, since the unbalance force varies with the speed squared, $(5,000/10,000)^2 = (\frac{1}{2})^2 = \frac{1}{4}$ or therefore one-fourth the force would allow four times the residual unbalance value or 4×16 gm.-in. $= 64$ gm.-in. (2.26 oz.-in.). The eccentricity is 2.26 oz.-in./(1000) lbs. (16 oz./lb.) $= 2.26$ oz.-in./$16 (10)^3$ oz. $= 0.141 \times 10^{-3}$ inches $= 0.141$ mils. The vibration expected would be 2×0.141 mils $= 0.282$ mils peak-to-peak.

The expression used above, oz.-in. $= F_u (1000)^2/1.77$ (rpm)^2 could have been stated as in API Standard 617 for compressors and Std. 612 for turbines, oz.-in. $= 56,347$ [Journal Wt. (lbs.)]/(rpm)^2 but this is the same amount.

This exercise was written to orient you to the fact that the balance tolerance divided by the rotor weight yields the eccentricity. Using ISO Standards (International Standards Organization), values will be expressed in ew units (mm-rad.)/sec. for quality of balance. I have used these values, along with API and Navy 4W/N values for 10 years and they correlate fairly closely with API; however, Navy and ISO Standards do not express the speed in squared terms which causes these values to deviate or cross if plotted over a wide speed range. It appears to me that ISO adopted the VDI Standards (Society of German Engineers) for the most part, then ANSI adopted ISO.

Comparing the examples just described to ISO Standards is helpful in understanding how to use the terms. A grade balance value of 2.5 (mm-rad)/sec. will be used, as suggested for turbine and compressors. Dividing

this term by rotor speed, in rad./sec., gives the eccentricity in mm. Multiplying this by the rotor weight in gms provides the residual unbalance allowed:

$N = 10,000$ rpm.

$w = 2\pi N/60 = 2\pi(10,000)/60 = 1046.6$ rad./sec.

$ew/w = e = 2.5$ mm/sec. (rad.)/1046.6 rad/sec. $= 2.388 \times 10^{-3}$ mm (94.04 microinches).

$U_r = 2.388 \times 10^{-3}$ mm $\times$ (1000 lbs.) (454 gm/lb.) $= 1084.46$ gm $=$ mm $= 42.70$/gm.-in. (compares to 64 gm-in. before) for 5000 rpm.

$U_r = 1084.46$ gm-mm $= 42.70$ gm-in. (compares to 16 gm-in. [API] before).

Using the ISO or ANSI balance tolerance charts, (see Fig. 8-14) plot speed on the absissa vs. the balance grade (diagonal line), read U_r and e on the ordinate.

Multiply this Residual Unbalance Value by the rotor weight to find the balance tolerance for acceptance.

Both the API and ISO limits have one common advantage, for use by an operating company's Mechanical Inspection Department. Only the operating speed and rotor or journal weight are necessary to inspect, say, a repaired rotor. Any guess work on quality, which can be inexpertly given by a maintenance supervisor, is removed. Residual unbalances which create, at speed, only 10 percent of the rotor's weight, at rest, simply cannot give you deleterious effects caused by excessive unbalance.

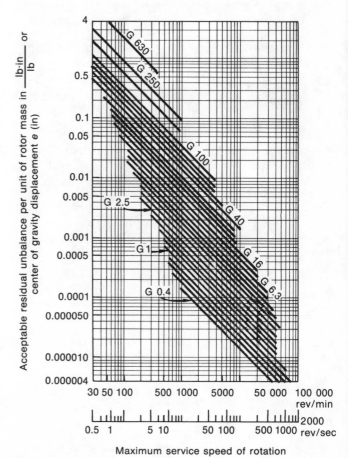

Fig. 8-14—Balance chart from ANSI $2.19, 1975 (ASA Std. 2, 1975).[3]

A sensitivity check on the balancing machine can be performed as outlined in API 617. This check simply assures that a balancing machine has the sensitivity to give you the acceptable limits. After completing the balance, if a series of multiple weights (equal to the balance tolerance) are placed on a rotor, the balancer readout should plot a fairly straight line to zero with some allowance for error.

From our first example, W_{BT} would be 16 gm-in. Using ½, 1, 2 and 4 times multiples of this weight, one develops 8, 16, 32 and 64 gm-in. weights. Placing these alternately on a rotor that is considered balanced should develop a plot such as shown in Fig. 8-15. (Note: Do not attempt this on an unbalanced rotor as all balancers exhibit non-linearities at high unbalance values.)

Placing weights on a rotor should not be a check on calibration but only on sensitivity. The balancing manufacturer's procedure for calibration should be followed. If the balancer uses a drive shaft coupling (typical of Schenck-Treble), the coupling drive should be reversed 180 degrees at balancing completion with no increase in residual unbalance more than double the accepted level. For example, if 8 gm-in. was determined as the balanced level, reversing the drive should not exceed 16 gm-in.

On the API data sheets for purchasing equipment, there is a place for specifying field balance provisions should equipment need trim balancing in service. This is the time to request such a provision for minimum cost, the proper design, number of balance holes, balance planes, arrangement, accessibility through casing, etc.

An attempt would probably not be made to trim balance more than 1/3g using field balance planes. More unbalance than this should require shop balancing. Also, the vibration level would normally force a shutdown.

Special tools are needed to do field balancing. In addition, a separate skill must be developed to determine the amount and provision for weight attachment. More important, proper selection of the trial weight placement is necessary to ensure that an improvement is made on trial runs. Special machined materials with approximately 83 percent tungsten are available with twice the densities of carbon steel [Hevimet (GE) and Kenturnium (Kenametal)].

Balancing programs. Several single-plane balancing techniques are described in Chapter 4. Further, single-plane and two-plane balancing programs for proximity probes with run-out consideration are listed in the Appendix.

These programs are prepared under the phase convention listed on the worksheets. This format records the phase angle of the high spot against rotor rotation from the proximity probe. Trial weight placements and the correction weight placement uses the identical convention. This convention was used since the phase meter,

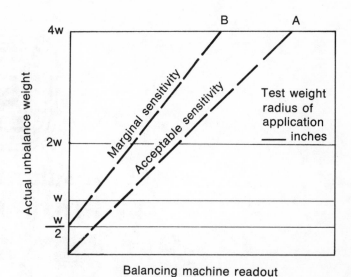

Fig. 8-15—API balance sensitivity chart.[5]

synchronous tracking filter and trim balancing instrumentation normally display the phase in this manner.

The programs were prepared for the H.P. Programmable Calculator, HP 67 (HP 97).

ACKNOWLEDGMENTS

The author is indebted to IRD (Messrs. John Harrell, Glen Thomas, Randy Fox and Pete Bernhard), Bently Nevada Corp. (Mr. Robert Eisenmann), Spectral Dynamics (Mr. Jon Palm), Dr. E. J. Gunter, Jr. (University of Virginia), Mr. V. Ray Dodd (Chevron USA), Texas A&M University (Mr. Eddie Morgan) and Monsanto Co. (Mr. Marvin Ringer) for assistance in verifying these programs. I hope the listing of these men explains the need to prove a system on real machines and real instruments.

Programs for strobe light systems are available from IRD for single plane and two plane balancing. Probe balance systems from IRD, however, should be restricted to their convention.

LITERATURE CITED

[1] *IRD Training Manual,* General Machinery Vibration Severity Chart, Form No. 305 D, IRD Mechanalysis, Columbus, Ohio.
[2] Blake, M. P., "Vibration standards for measurement," *Hydrocarbon Processing,* January, 1964, Fig. 1.
[3] Acoustical Society of America, New York.
[4] Rotoflow Corp.
[5] American Petroleum Institute, New York.

Alignment by Proximity Probes

The proper alignment of rotating equipment is very important to successful and continued operation. More than half of the contributing faults towards poor performance determined during several years of vibration diagnostic analysis, was attributed to misalignment. There are various types of drive couplings for turbomachinery trains; yet, all these devices provide some degree of decoupling ability for the two shafts joined together by the coupling. A coupling design might claim $\frac{1}{2}°$ of pure offset. This represents 8 mils per inch of gear separation (87 micro-meters per centimeter of length) before the gears in the spline would lock. Surely this is going too far and one should align to much closer tolerances. It is preferred to be within $\frac{1}{2}$ mil per inch of gear separation (5 μ M/cm).

An improper misalignment will be indicated by a twice running speed frequency with increases in normal running frequency plus axial vibration values approaching that of the radial vibration. The 2x has always been noted for misalignment between two machine casings, and the axial percentage of radial vibration can vary by each case but becomes highly significant at values equal to the radial (worst radial) vibration amount.

The negative results from misalignment are:

1. Extreme heat in couplings

2. Extreme wear in gear couplings and fatigue in dry element couplings

3. Cracked shafts and totally failed shafts with failure due to reverse bending fatigue transverse to the shaft axis initiating at the change of section between the big end of the coupling hub taper and the shaft

4. Preload on bearings (This will be evident by an elliptical and flattened orbit resembling a deflated beach ball. Pure asymmetry of vertical and horizontal vibration can be misleading since the bearing spring constants could vary greatly in the k_{yy} (vertical) and the k_{xx} (horizontal) axis.)

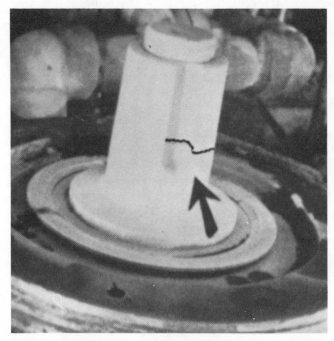

Fig. 9-1—Cracked shaft on a steam turbine.

Fig. 9-2—Gear tooth coupling after three weeks at severe misalignment (over 80 mils).

Fig. 9-3—Steam turbine inlet with detail of tow eddy current probes at one corner, steam end left side.

5. Bearing failures plus thrust transmission through the coupling which can be totally locked (Axial vibration checks across the coupling, i.e., at each adjacent machine will generally confirm this condition.)

There is at least one positive result from misalignment, i.e., moderate misalignment, and that is to restrain oil whirl. Since oil whirl is a condition that can result due to high speeds and light loads, misalignment has been effective, in small percentages, to eliminated whirl.

There are several techniques for measuring both cold and hot alignment. The procedure outlined here for cold alignment will be by reverse dial indicators and graphically plotted on ten division graph paper. The hot measure will be using Bently eddy current proximity probes in the 300 mil (7.6mm) tip size mounted in 3/8-inch (9.525mm) diameter smooth unthreaded sleeves. The sleeves are 4 inches in length (10.16 cm) with the 95 ohm coaxial cable extending one meter (39.37 inches) to the first connector. The extension cable will be 3.5 meters (138 inches) and connected to a proximitor of the 7000 series and selected for 100 mv/mil normally for long range measure. However, the 200 mv/mil sensitivity can surely be used provided the expected movements would not exceed the usable calibration range.

Nozzle load check. It is also important to note that the stands (holding probes) can be used very effectively to determine the movements of a drive steam turbine, for example, when the steam headers are pressured to the trip valve. Pressuring the steam inlet headers to the trip valve and the exhaust or extraction lines to their respective block valves is an early check on nozzle loads by the piping and should be conducted at the earliest oppor-

tunity. Early checks for piping changes can remove a severe and timely delay to a start-up schedule of a new operating plant. Movements at the four corners of a drive turbine within 5 mils (127 μ M) is considered acceptable. During this test, no heat, e.g., steam, is applied to the steam turbine, only the piping.

Alignment example. The simplest way to explain an alignment procedure is to walk through an actual step by step procedure. A two-case train will be used consisting of a steam turbine drive through an 18″ spacer cou-

Fig. 9-4—Turbine and compressor.

pling (Bendix) to an air compressor. The speed will be 4,400 rpm and the horsepower at 10,000.

Target installation. This problem of alignment really starts at the mechanical specification stage. Here, alignment targets were requested by the equipment builders for four "L" shaped alignment targets as shown in Fig. 9-5. One each of these targets to be mounted on all four corners of the equipment at the bearing housings, i.e., near the shafts, and the supports should the bearing housings be inaccessible. An alternate target is shown in Fig. 9-6 for retrofitting in the field on existing equipment. The smaller target can be installed simply using a spot face tool with pilot, as available from any tool supplier. A single machine dowel at $3/8''$ (9.525mm) diameter and one diameter depth in penetration is required. No attempt is made for an interference fit but rather a slip fit using an adhesive. In this case, Loctite 35 with primer. For installations on flat surfaces, a belt grinder, or equivalent, satisfies that requirement. A Bayflex hand grinder does a good job.

Sole plate and adjustments bolts. The builder was also requested to provide sole plates at 10″ (25.4 cm)

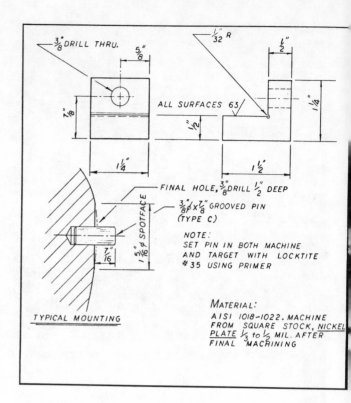

Fig. 9-6—"L" shaped target for spot facing, dowel, and epoxy.

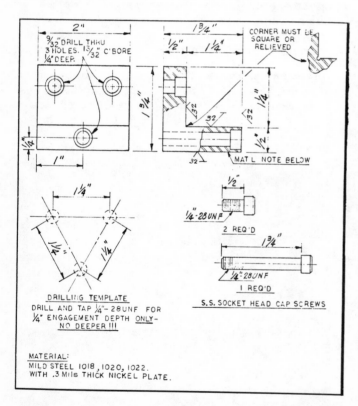

Fig. 9-5—"L" shaped target for bolting and cement.

larger size on the three outer perimeters. This larger size has a two-fold advantage. First, it allows multiple possibilities in locating the water cooled stands which hold the proximity probes. On final assembly of all the machine components, there are many interferring pipes and conduits, plus bracing in the way. Secondly, this encourages the builder to be more practical on sizing alignment bolts for adjusting the equipment in the field. API 612 and 617 do require adjustment bolts but the purchaser should ensure that x, y, and z directions are allowed, i.e., horizontal, vertical, and axial.

Shims. The installation procedures which often start with the equipment builders[1] general arrangement drawings should specify stainless steel shims at 125 mils (0.32 cm) as a minimum under all equipment supports. This would include the fixed supports of the wobble plates (flexible plates) of a turbine or compressor. It is preferred that this shim be one machined plate. Should multiple trains of duplicate equipment be included using dry element couplings, then this is a good time to request shims for the coupling flanges to assure that proper interchangeability exists on each machine and spare rotor and spare coupling.

The balance of instrumentation and tooling need only be supplied by the owner, contractor, or service group performing the alignment.

Fig. 9-7—Sole plate alignment bolts with target, stand, and probes.

"L" targets. The "L" targets are machined with a 1/3 - 1/2 mil (8-12 μ M) plating of nickel over the base mild steel stock. This protects the target from the environment, is inexpensive, and allows inductive calibration or magnetic holders to be used without harm. The "L" shape is also practical should an optical reference or optical alignment tooling be necessary at some later date. It can be seen in Fig. 9-9 that dual optical scales can be held off a single magnetic base by applying a little ingenuity in ordering or modifying standard magnets.

Targets should be mounted at the centerline of equipment and in the lower half of axial (horizontally) split casings.

Water cooled stands. The reference stands should be simply understood. A stable reference is needed to observe the movement of a piece of equipment in two directions (horizontal and vertical). If the shaft can be measured, then by all means, measure the shaft. If not, measure the bearing housing and lastly the case, which houses the bearing housing, which houses the shaft. Simply stated, measure the shaft or measure what tells the shaft where to go.

Back to the reference stands, a stable point is needed to hold some proximity probes in two attitudes, vertical and horizontal, to plot Cartesian coordinates of the machine movements during the rise to equilibrium temperatures

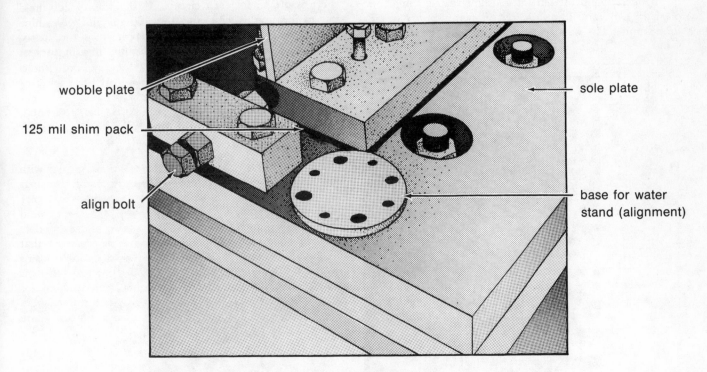

wobble plate

125 mil shim pack

align bolt

sole plate

base for water stand (alignment)

Fig. 9-8—Compressor wobble plate with 125 mil shim pack.

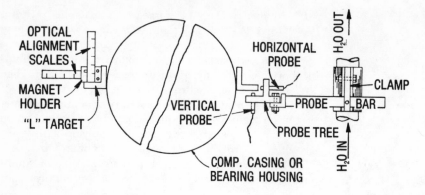

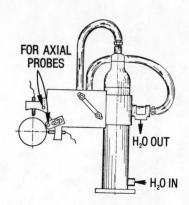

Fig. 9-9—"L" targets used at bearing housing or casing allow optical or proximity hot alignment measures.

Fig. 9-10—Typical water cooled reference stand for hot alignment measures directly on shafting simultaneously measured across the coupling using proximity sensors.

and mass flow rates. The water cooled pipe stands, with brackets, do that job well. First, the pipe should be 2″ in size and flanged at the base. The flanged end of the pipe should be sealed to prevent water leakage. It is preferred to provide matching blind flanges which will be tack welded to the soleplates; however, the pipe flange can be oriented and bolted to the soleplates. The upper end of the pipe should be swaged to ¾″ or ½″ NPT male for pipe fitting with a 180° turn to vertically down with a standard garden hose fitting. Water will flow through the pipe at all times keeping the reference stand from thermally growing from heat of the rotating equipment or any nearby systems, hot or cold.

It is interesting that cooling tower water heat loads in an operating plant remain constant through night or day and will vary little in water temperature, say 87° F for example. The water can flow in parallel through each stand or in series, i.e., alternately. It is often easier to

connect the water stands down one side of the train in one series hook-up and the opposite side of the stand in the same fashion.

In taking data, one should observe that the water stays on and not stopped by a conscientious energy saving operator. Small installations (pumps and drivers) may dictate 1½ inch water cooled stands as the space would be more limited.

Calibration. Calibration of the probes should be performed on the actual targets used or the same shaft material if shaft direct readings are to be taken. A section of the target material can be cut from one of the targets and mounted on a micrometer pin for calibration. A curve is drawn for each circuit. A 100, 150, or 200 mv/mil calibration is best and the 100 mv/mil gives the longest range. It is easier to interpret since the digits help, i.e., 0.1 volts equals 1.0 mil. Connecting the probes

Fig. 9-11—Two examples of water cooled stand bracket across couplings.

Fig. 9-12—Measurement stands down the compressor train-steam end to compressor discharge.

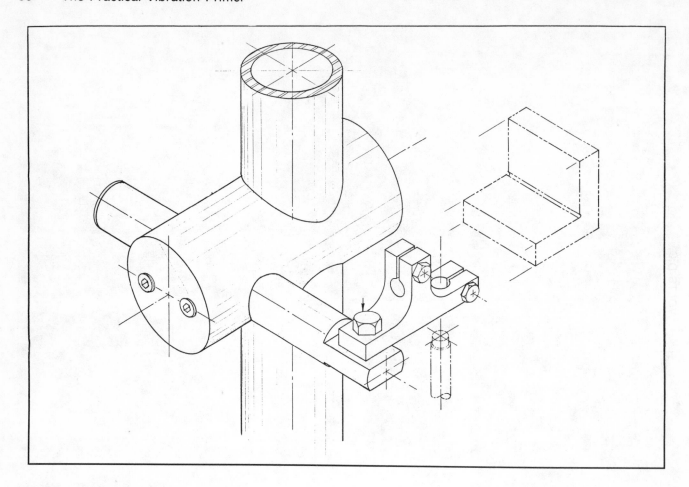

Fig. 9-13—Water cooled stand for "L" target.

to the holders and rough gapping with a feeler gauge helps during the mechanical setting of the probes. It is easier to set the gap needle using the voltage reading. For example, if 90 mils equals 9.0 volts, then the probe can be adjusted to 9.0 volts on the voltmeter very easily.

Note: When taking data on the shaft direct during hot alignment runs, it is more correct to note the change in readings when the shaft first starts to roll and make a slight correction for shaft run-out. This correction could be plus, minus, or not worth correcting. The probe reading a shaft direct gives you the average of the shaft run-out. A square shaft could be measured, if necessary, without loss of accuracy. (It should be apparent that no correction is needed when recording from case or bearing cap targets.)

Log sheets. A log sheet is needed for taking data and noting changes in operation such as increased load, speed, discharge temperature, oil temperature, etc. The log sheet should be systematic, and each channel of instrumentation should be labeled against a probe number and position.

A sample sheet for taking data is shown in Fig. 9-15. A further aid is connecting simple DC voltage recorders, typically Rustrak, to record data during the entire run

plus capturing all transcients in temperature which is impossible when logging data. The voltage range of the recorder should be 20 VDC with convenient divisions and a tape speed of one foot per hour is convenient. This would give only 8 feet per shift or generally 24 feet per total run per point. The recorder is valuable when recording shaft axial thermal transcients. Rustrak also provides a hand roller wind-up assembly for reviewing tapes. The tapes are pressure sensitive so no ink or heat is required.

Instrumentation. The instrumentation is simple. Either four or five channels is grouped in one fiberglass case. The case has provisions for leads coming into the box with the lid closed so that the lid can be closed when not in use or during inclement weather. A plastic transparent bag is also used over the equipment during rain periods and the data can still be recorded looking thru the plastic (with a flashlight at night).

A digital voltmeter is provided to assist in calibration. However, readings by fixed meters and by DVM should be recorded initially before heat is applied to the system. It is more convenient to read direct voltmeters. Should a meter malfunction for any reason, then the DVM readings can be used.

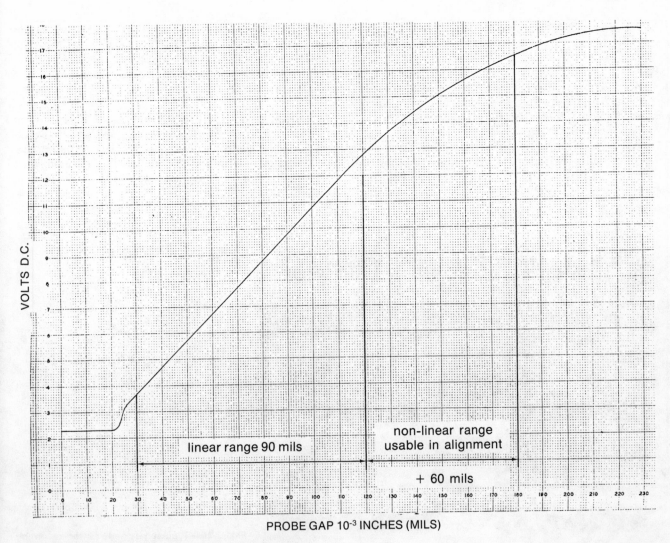

Fig. 9-14—Calibration set-up plus the resulting curve.

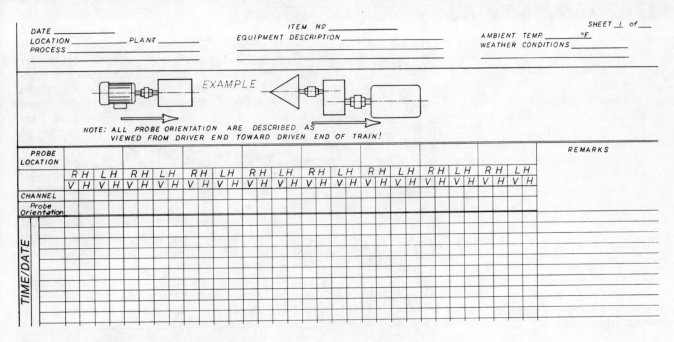

Fig. 9-15—Log sheet for recording data.

A jack is provided for recorder hook-up and recorders with 100K ohms or greater input impedance should be specified.

Power is available by AC with back-up power on DC should AC not be available. Power cord connectors should be protected and turned on their side during inclement weather. Care to use intrinsically safe electronics is important and not restraining.

Dual scale meters are available reading directly in mils rather than volts, but this involves another mechanism to fault on portable equipment and generally is not desired as an additional feature.

Coaxial connectors should be wrapped with Teflon tape with an outer restraint using vinyl tape. This will prevent any ill effects from wind, dirt, or rain (or steam).

Typical read-out boxes and recorder boxes are shown in Figs. 9-16 and 9-17.

Graphical plot. The operating train is laid out on ten division graph (or chart) paper with one division equaling 1 inch in length. Transverse to that layout, the vertical or horizontal movements are laid out with 1 division equal to 1 mil. Thus the chart, while linear, is amplified in movement by 1000:1 scale.

Fig. 9-16—Typical recorders for recording changes with time.

Fig. 9-17—A five-channel alignment box for recording data.

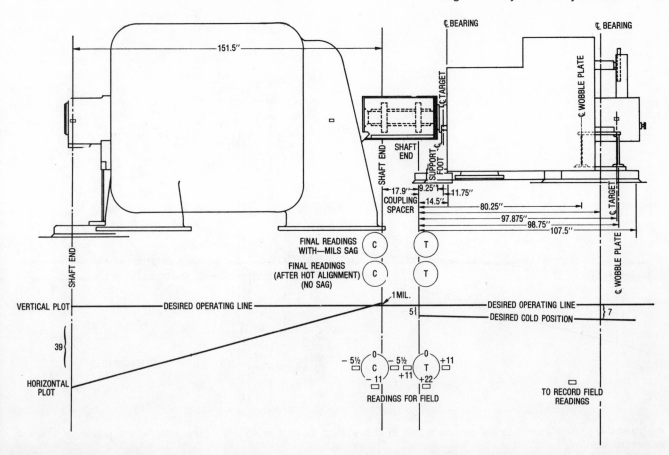

Fig. 9-18—Alignment Plot 1 shows the first layout of known information for a turbine/compressor train based on the manufacturer's data for heat rises.

Important parts of the operating train are position in this layout.

- The casing supports where shimming is performed

- The shaft ends and the coupling spacer length

- The position and span of the reverse indicators to be used

- The location of targets where measurements are to be taken

- Bearing positions, only because machinery manufacturers may give rise data at this location

Plot 1 (Fig. 9-18) shows the turbine and compressor laid out with the manufacturer's predicted heat rises. In this example, the heat rise at the steam end bearing is 7 mils; at the exhaust shaft end, 5 mils; at the compressor coupling (suction) end, −1 mil; at the compressor outboard shaft end (discharge), 39 mils. Since three out of four of these points are heat rises, they are plotted below the hot operating line for a desired cold position. The inboard compressor end having a minus or drop amount is plotted above the hot operating line. The expected horizontal shift is zero and can be plotted now or later.

The required reverse indicator reading can be taken from this information to present to the field. This is given here in two forms. One is the true reading assuming no sag in the indicator bar. The other assumes, in this case, a one mil sag in the bar spanning the coupling gap. The appendix of this procedure has some examples of misalignment versus indicator readings plus an example on developing readings from the plot and also plotting from the readings. With some study, basic facts come forth which are easy to remember:

- Halve the indicator reading to determine shaft center shift for plotting, e.g., a "0" to "+ 12" reading indicates the shaft center is 6 (12/2) mils out or away from the center of rotation of the indicator bar mounted shaft being turned.

- Double the shaft centers from the plot to determine what the reverse indicator will read.

- Sag in the indicator bar presents an error that must be corrected and the error is twice the actual sag, i.e., a 2 mil sag causes a 4 mil difference in reading on verticals and the same 2 mil error on horizontals. A minus reading will increase with sag; a plus reading will reduce with sag.

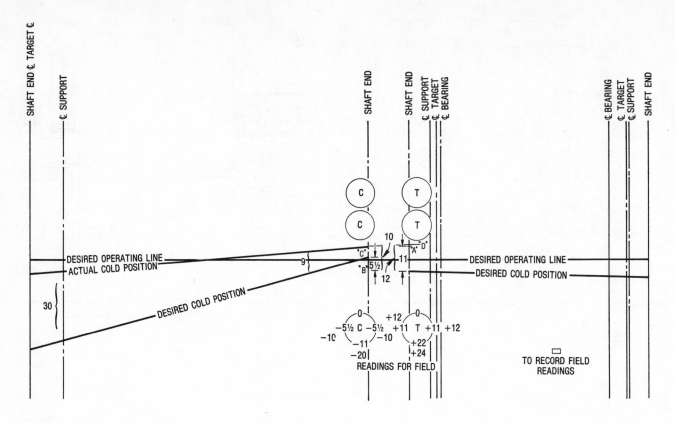

Fig. 9-19—Alignment Plot 2 shows the advantage of this system in positioning a machine from where it lays to where it is needed cold.

- Sag must be added when taken from the plot. Sag must be subtracted when taken from the field readings.

- Pushing in on an indicator reads "plus" or "right." Pulling out or reaching of the indicator stem causes a "minus" or "left" indicator.

Indicators. Indicators have been specially selected from Brown & Sharpe (Rhode Island) to have center travel positions with revolutions count wheels in two colors for a total travel of 1 inch (± 500 mils) to prevent error in reading the indicator.

The indicators can be mounted on the equipment shaft rather than the indicator bar to remove the weight of the indicator and thereby reduce sag errors.

Positioning the machines cold. One of the great expedients in this technique is shown in Plot 2 (Fig. 9-19). The field measurement has now been determined and shims are necessary to put the compressor shaft in the correct cold position. It was determined on this job to move the compressor rather than the turbine which is generally the easier move. By plotting the reading taken in the field, the compressor shaft relative to the turbine shaft is determined. By measuring the number of divisions at each compressor support (divs. × 1 = mils), the correct shim change can be determined. In Plot 2, the correction is − 18 mils at the discharge end and + 7 mils at

the suction or inboard end. This can take days by trial and error. Should large corrections be needed, the exact improvement may not be obtained. However, it should be safe to run the hot alignment run from this position. The actual data will generally require adjustment anyway.

Final corrections. Once the equipment has been operated at design conditions for a certain period of time, the actual measurements are recorded. This has to be the best information for final adjustments. The heat rise equilibrium temperatures can be considered as terminated when the readings do not change or start a slow vacillation over an hour or two.

Plot 3 (Fig. 9-20) is simply a reconstruction of Plot 1 except that the measured data is plotted at the target points and a new cold position (final) is determined.

The equipment must be allowed to cool totally, i.e., steam blocked and bled on the turbine (sometimes a purge of nitrogen or dry air thru the turbine can speed that delay).

Note: The condition of the warm lube oil must be consistent in all the cold measurements. It is suggested that a machine at "rest cold" should have:

- Steam pressured up to the trip/throttle valve

- Exhaust steam blocked at the header valve

- Turbine drains open

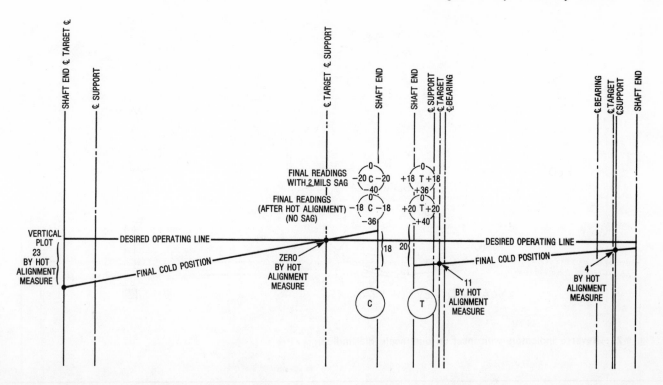

Fig. 9-20—Alignment Plot 3 shows the revisions to Plot 1 after hot alignment data are obtained. Plot 3 would be identical to Plot 1 if heat rise predictions were accurate.

- Lube oil circulating at design inlet temperature, i.e., 110° F

- Compressor blocked in suction and discharge

Lube oil on or off can vary the initial heat rise by four to six mils during winter weather.

The final plot should be maintained in the maintenance file for that equipment. Any removal later for repairs should be reinstalled for the same position even though the machine may have shifted due to foundation settlings, shim growth, loose supports, etc.

The final corrections should be made from the final reverse indicator readings particularly if it was the first time operated. A machine will rarely return to the start point on the first load and heat cycle.

It is further recommended that on new equipment alignment, a four hour demonstration run be completed successfully and allowed to cool before the hot alignment run is started. In this way, the various mechanical problems can be corrected and the alignment run will not be interrupted. Further, it gets all the extra people out of the way—machinists, insulators, instrument people, by-standers, and pipe fitters. The machine should be roped off to prevent someone from crawling over the unit and bumping everything out of position. The mechanical problems mentioned previously have proved to be oil leaks, steam leaks, trip valves unlatching, governor failing to take over, governor running rough or swinging, tachometer not working, seal rubs or bearings running hot.

Special systems. Axial growth of shaft ends can easily be performed by adding two probes reading off the coupling hubs for dry membrane couplings. The dry coupling is significant to this measure in the first place since many couplings can stand only 62 or 125 mils of axial travel as a limit.

When a gear is involved in any system, the gear becomes the reference. Everything references to and is aligned to the gear. Further, the high speed end references to the gear pinion shaft. The low speed end references to the gear shaft (low speed). Gears are usually aligned at the casing to ensure good gear mesh tooth contact. This type of alignment is very important and should not be disturbed to perform a train alignment. This again emphasizes the need for a minimum of 125 mils of shim stock under all feet during machinery installation.

Reference to other aligning systems. There are other aligning systems including lasers, optics, Dyn-Align bars, mercury levels, and the characteristic "shut-down-and-hot-check." There are other face and rim methods of shaft aligning. The reverse indicator method is felt to have the overall advantages against the errors of shaft float, lowest span percentage to foot span, surface defects (metal), and difficulty in display. The hot aligning tools require about the same overall outlay of capital. Each technique has advantages and disadvantages. The Dyn-Align bars have definite advantages on shipboard alignment where a basic reference is difficult. The optic system is easier to set up and take down, but it cannot be

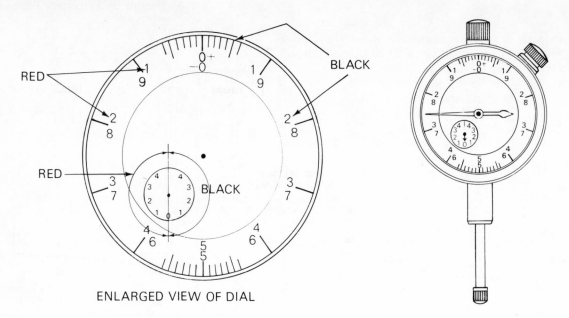

RED

BLACK

RED

BLACK

ENLARGED VIEW OF DIAL

Fig. 9-21—Reverse indicators with inset on dual scale dial indicators.

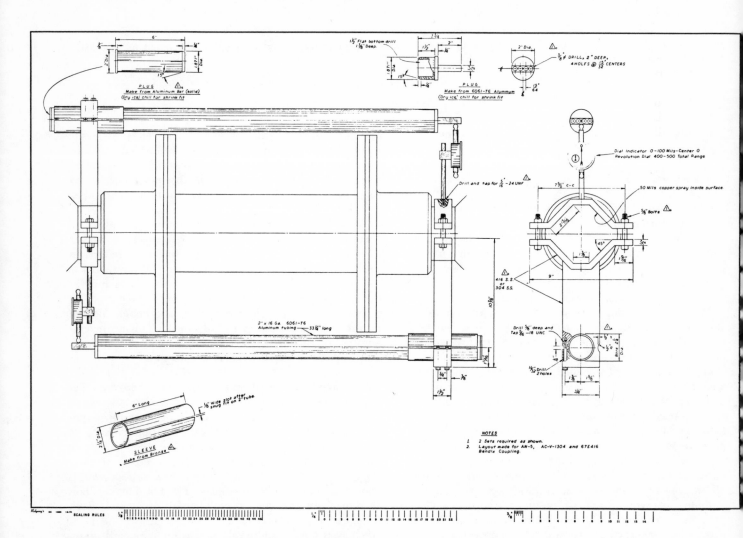

Fig. 9-22—Reverse indicators and low sag bar in place.

read continuously in poor weather nor read quickly. The laser has less advantages. Readings must be made at the target and it is not very reliable.

The water stand/probe procedure outlined here is felt to have the best overall accuracy, understanding, and information. In summary, the advantages should be reviewed:

1. Water pipe stands are stable references.

2. Two movements are plotted at each selected position.

3. The measurement locating can quickly be custom fitted to each job.

4. Stands, probes, holders, leads, and instruments can be reused on succeeding jobs. Possibly some new stands may be required due to heights.

5. Extreme flexibility is provided for positioning probes.

6. Proper aligning of probe-to-target or probe-to-shaft is assured individually at each single position.

7. Measurements are capable in very crowded piping clusters (often the downfall of optics).

8. Measurements can be made over extended time periods without difficulty, during poor weather, and allowing recording of data.

9. It is the only technique to measure direct to the shafting.

10. It is the only technique to measure directly axial transcient shaft growth.

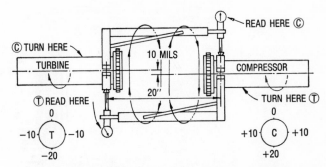

Example 2—Reverse indicator readings with the right side shaft 10 mils low of left shaft and zero angularity misalignment.

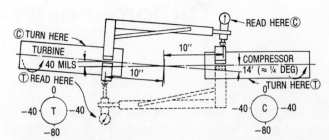

Example 3—Error in true offsets of Example 2 caused by a 2 mil sag in the indicator bar.

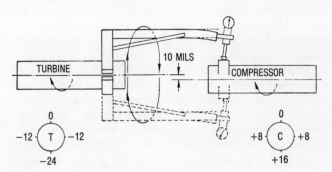

Example 4—Reverse readings taken with symmetrical angular offsets.

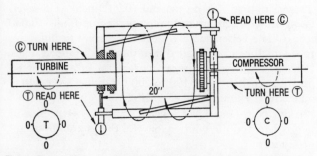

Example 1—This illustrates the readings obtained with zero misalignment and zero sag in the indicator bar.

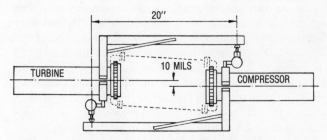

Example 5—One technique for reducing indicator bar sag. Indicator weight is held by machinery shafting.

How to Prevent Turbomachinery Thrust Failures

Machines such as turbines and centrifugal compressors can be severely damaged if the thrust bearing fails unexpectedly. Failure can easily cost over $1/2 million including lost production. Such a failure can be prevented if careful consideration is given to the machinery and its protective instrumentation.

Most machines of this type have a thrust bearing of the hydrodynamic (slider bearing) type which develops an oil wedge lifting the thrust collar or runner from a stationary bearing, which could be of the straight flange, tapered land or tilting pad type. Actually, a thrust bearing is really an assembly which has two bearings. One bearing takes the normal or expected thrust loads imposed by the rotor. The other bearing would therefore be located, in design, to take abnormal or unexpected thrust loads. Many vendors will refer to one as the active (normal) thrust bearing and the inactive (counter or abnormal) thrust bearing.

Because of thermal expansion of the bearing elements, the two bearings could grow into one another causing severe loading and immediate failure. For this reason, the contact on each bearing is separated a certain distance based on the operating environment temperature. This separation distance is often referred to as the float zone between active and inactive thrust shoes (pads, plates, surfaces). For steam turbines, this distance might be 0.009 inches up to 0.014 inches. Thrust bearings for steam turbines will normally be at the steam inlet end since the rotor will expand from the thrust bearing. The blading (rotor) should move away from the nozzles and stationary blading as this is a critical and sensitive dimension.

On the other hand, centrifugal compressors are more lenient, with cold rotor floats of 0.015 inches to 0.022 inches typically. Total rotor float of a centrifugal compressor (with thrust bearings removed) moving axially in the case from one staging to the next could be in the order of 1/4 inch (0.25 inches).

There is nothing unusual when a machine, in operation, floats randomly between the active and inactive bearings. I would consider this quite healthy exposing basically no thrust bearing loads on the bearings, provided an axial vibration is not set up which could cause fatigue or efficiency swings.

Thrust loads can be measured in many ways: Bearing metal temperatures, oil exit temperatures or heat differences, load cells behind the pads or bearings, or by deflection of the pads. Temperature and load measurement are recommended for actual engineering data or load distribution for attitude and alignment studies. Rotor deflection in an axial direction is recommended for machine protection monitoring with automatic shutdown to prevent severe failure damage.

Thrust failure protection is the topic of this discussion, and a systematic procedure is outlined with case histories of "saves" from the worst type of machine failure.

Different types of sensors can be used on a continuous basis to measure thrust movement or deflections. Earlier developments used both air and hydraulic nozzles which varied in back pressure as the distance between the nozzle and a part on the rotor (often the thrust collar or trim balance ring) changed. Calibrations could be obtained with sensitivities of 1 psi per mil (0.001 inch). However, sensing ranges were often limited to 25 to 30 mils (0.030 inches). The better sensor is the eddy current or inductive probe which can be installed in rather small tapped holes, 1/4 to 3/8 inch, and can sense movements in excess of 80 mils with sensitivities of 100 to 200 mv/mil. (See calibration curve at 100 mv/mil—Fig. 10-2.)

At this point, one must develop a proper perspective of the problem and the solution. Several things are important to consider:

• Thrust bearings are known to fail in about 30 seconds; yet, immediately prior to failure the bearing on examination may have zero defects until the supporting oil film breaks down.

• Cost of the thrust bearings may be $700 or higher for pads only and may be $5,000 for a total assembly. A machine wreck by thrust bearing failure will intermarry the rotating parts to the nonrotating parts with a repair cost

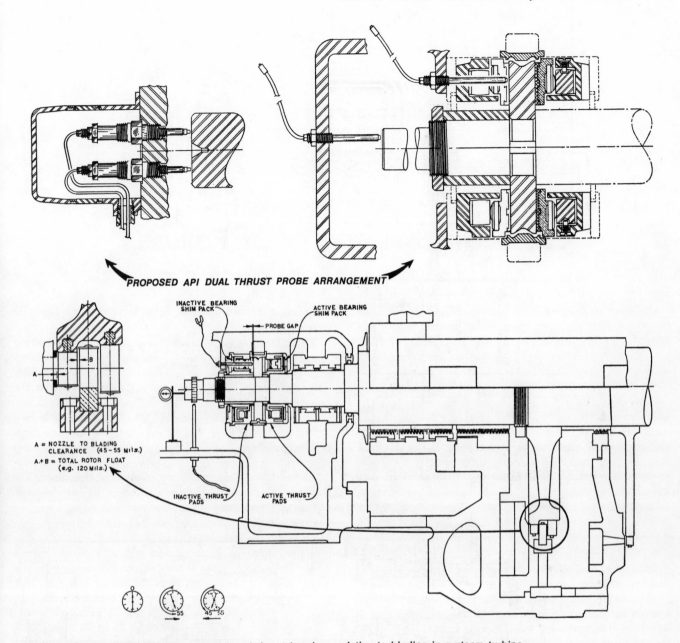

Fig. 10-1—Dual thrust probe arrangement and thrust bearings relative to blading in a steam turbine.

of $1/2 million or more. Therefore, the thrust bearing (failed) should be sacrificial to the big machine.

● Trying to save the thrust bearing will only cause severe confusion. First, if shutdown occurs before babbitt failure, inspection will indicate the thrust bearing was okay, but the sensor and related instrumentation indicated the onset of a failure! Unnecessary shutdowns will lead to a disarming of the shutdown system rendering the system incapable of proper protection. Second, the thrust bearing, being constructed of pads, for example, with self-leveling articulating links, and possibly shim packs of ⅛ to ¼ inch, can deflect 4 to 8 mils under load increases well within the design of a thrust bearing. Large deflections can be reduced and downtime minimized by machining single element shim plates to replace large packs of thin shim.

● Machine builders are naturally going to set short limits of rotor movement in their instructions for thrust shutdown instrumentation to protect their machine designs from possible criticisms. The builder doesn't care how many times you may shut down for inspections on movements of 5-6 mils (his suggested limits, but within expected deflections for rated loads).

● Limits for shutdown should definitely indicate that babbitt is being removed yet the shutdown response must be sufficient to bring the machine to rest within the babbitt thickness and definitely within the rotor travel clearance before contact occurs.

● A sensor system should be correlated with normal bump checks of rotors and dial indicator measurements.

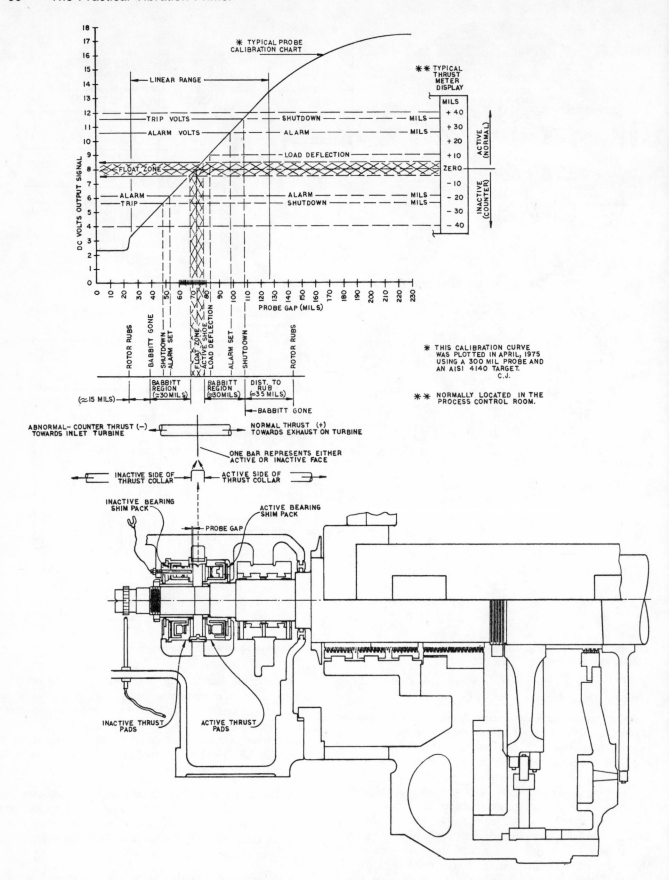

Fig. 10-2—Thrust bearing float and probe gap projected through calibration curve to rotor position readout instrumentation.

Fig. 10-3—Failed thrust bearing due to improper venting of buttons supporting thrust bearing pad. This is typical of a failure to properly bond temperature sensor to bearing babbitt (7,600 hp @ 6,000 rpm).

Fig. 10-4—Failed thrust bearing on compressor due to overload. Trip-out occurred manually at 17 mils axial deflection. Eight to 10 mils of babbitt removed by failure (5,000 hp @ 11,000 rpm).

This is a very good way to improve everyone's confidence in the machinery protection instrument system.

THRUST BEARING FITTING PROCEDURES AND SENSOR POSITIONING

Steam turbine example:

Specified thrust float by manufacturer = 0.009 inches − 0.012 inches.

Specified shroud (rotor blading) standoff to nozzle ring (closest contact) = 0.045 inches − 0.055 inches.

Specified total rotor float, minimum = 0.120 inches.

Procedure (see Fig. 10-1):

• With thrust bearing removed, move rotor until shrouds contact nozzle ring ("A" clearance = 0). Indicate with dial indicator (compressed more than 200 mils) and set indicator to read zero. Move rotor towards exhaust until first stationary blading stops the rotor on axial travel contact. Record this travel on maintenance setup sheet after assuring that greater than 120 mils was obtained (example specification from above).

• Return rotor to the zero reading (against the nozzle block), install only the active thrust bearings with shims. Move rotor towards exhaust until contact is made with the thrust bearings (lower half only is used at this point). If indicator reads 0.055 inches good; if greater, increase active bearing shim thickness until 55 mils is obtained. Place equal amount of shims in upper half active thrust bearing. If less than 55 mils, reduce shims until 55 mils travel is obtained.

• Place inactive thrust bearings with shims into place. With the rotor against the active thrust bearing, move the rotor towards the nozzle block shimming for a 46 to 43 reading. This yields a float zone of 9 to 12 mils.

This procedure could have been reversed, i.e., setting the inactive bearing shims for 45 mils standoff and then the active shims for 9 to 12 mils float. With a large float tolerance, this would be preferred.

Confirmation of this total float by a bump check with the top half bearings in place is necessary. Lack of confirmation, e.g., less travel, can be a real problem in troubleshooting machine fits and will not be covered here. This would fall within a maintenance paper concentrated at setting thrust bearings.

Sensor placement. A calibration curve (Fig. (10-2) is plotted for the eddy current sensor and its related readout device. This calibration curve will be for the metal of the shaft to be sensed and the plot will be volts (ordinate) versus gap (abscissa). Gap implies the space between the probe (eddy current sensor face) and the metal used in calibration.

The calibration curve is at a sensitivity of 100 mv/mil (Fig. 10-2) with over 80 mils of linear range, and the voltage at the center of the linear range is 8 volts. At + 40 mils from the center of liner range, the output voltage is 12 volts. At − 40 mils from the center, the voltage is 4 volts. (Note: At 5 mils on either side of the center, the voltage is 8.5 volts and 7.5 volts, respectively.)

For further convenience, assume the bump checks above yielded 45 mils to 55 mils or exactly 10 mils float.

Fig. 10-5—Failed thrust bearing due to overload but primarily from cavitation caused by thrust reversals of compressor operating in a surge condition for over 20 minutes. Automatic trip-out occurred at 28 mils axial deflection. Sixteen mils of babbit removed by failure.

Because the center of the calibration range of the sensor system from the calibration curve occurred at 73 mils gap with an output voltage of 8 volts, this point will be used for the center of float position. Two methods can be used to secure this position:

1. Set the rotor in the center of the float zone by dial indicator and adjust the sensor to thrust collar (or shaft end) gap until 8 volts output is obtained at the thrust monitor output jack. The zero adjust would manually be moved until the readout meter indicates zero (center of scale). Bump the rotor to confirm both indicator and voltage agreement, i.e., -5 mils $= 7.5$ v, zero $= 8$ v, $+5$ mils $= 8.5$ v.

2. Move the rotor against the active thrust shoes and adjust the sensor gap until a voltage that corresponds to 5 mils active direction is obtained at the thrust monitor output jack, e.g., 8.5 volts. The zero adjustment calibration would manually be moved until the readout meter indicates 5 mils (active). A bump check should indicate 10 mils of travel on the indicator and 10 mils ($+5$ mils to -5 mils) on the readout meter of the sensor system.

Either method will work; however, the author prefers the last method with present instrumentation. It is easier to hold the rotor against one end of the travel. Further, should you wish to set up all machines against the active bearing and have the same meter reading regardless of float zone; then Method 2 (above) becomes a standard practice.

Most turbines will require that the sensor be placed on the steam inlet shaft-end or at least outside (outboard) the thrust collar. Further, the thrust will normally be toward the exhaust so that on turbines an increasing probe gap will indicate a movement in the active load direction. Conversely, a decreasing gap will indicate an inactive (counter) movement. One arrangement of shaft-end (near thrust bearing) dual probes used for thrust movement

sensing is shown in Fig. 10-1 as taken from a current task force effort of the API Subcommittee of Refining Equipment. A new specification, API 670, was recently adopted as a complete specification for vibration and thrust movement monitors.

Because a compressor's normal thrust can easily be in either a gap decreasing or increasing direction, it seems normal to use the turbine direction, calibration, polarity, etc., as standard and call out the compressor system by specific design.

It is worth noting that new instrument designs may soon require limited adjustment of the zero meter indicator in favor of self-centering zero meter convention. This will commit Method 1 to be the preferred method and give equal travel range to active and inactive movements of the rotor. This presents no problem to the user. However, to meet the requirements of all current types of turbomachinery, it shall require a minimum of 80 mils linear range by the sensor system builder.

Centrifugal compressor. Centrifugal compressor thrust bearing shims are set up similarly. However, the compressor rotor may have 250 mils of total float and will be set up in the center of float with a spacer or shims to give 15-20 mils (typically) of float. Efficiency of the compressor and possible aerodynamic initiated instability would be adversely affected by sizeable excursions from this position.

The operating, maintenance and diagnostic people have found that it helps to place two strips of black tape on the thrust meters to indicate the limits of the free float positions when a machine has been calibrated and checked.

THRUST ALARM AND SHUTDOWN LIMITS

Based on more than 20 years of experience with thrust bearing failures (over 50 by count) and use of various

Fig. 10-6—Failed steam turbine thrust bearing due to overload from selective fourth and fifth stage salt deposition on blading. Salting caused by impure boiler feed water (15,000 hp @ 4,500 rpm).

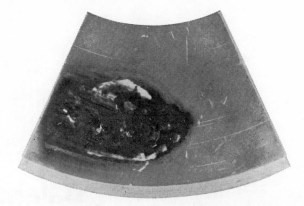

Fig. 10-7—Failed expander thrust bearing (one pad only) after 4,000 hours operation. Failure caused by conversion to synthetic lubricating oil with higher heat generated because of oil's different specific heat.

thrust indicators from "squallers" (electric touch method) to hydraulic back pressure indicators for gap, to pneumatic nozzles[1] with back pressures versus gap, to present eddy current sensor systems; our limits have been well defined, and countless *saves* have been accumulated.

Thrust *alarms* have been set up as 15 mils from the normal commissioning rotor position at design conditions (generally with good practice this will be within 5 mils of the active bump position). However, one can exert little force on a thrust bearing during the process of manual prying to bump the rotor.

Thrust *shutdowns* have been set up as 5 mils past the alarm point or 20 mils from normal load deflection points.

Given the proper instrument system, this shutdown point can be extended 5 mils for a 25-mil set point value. At present, this has not been done with existing instrumentation systems. New instrumentation systems allow sufficient range (API 670).

Examples of thrust failures are shown in the photos (Figs. 10-3—10-7). These bearings while partially failed did not cause rotor damage, with the exception of Fig. 10-6. This failure did rub all trailing edges of each of 10 blading shrouds with repair allowed within one week. This failure occurred over approximately 2½ minutes (a true exception) and though the horsepower was 15,000, the speed was only 4,500 rpm.

Total thrust bearing failures generally show nothing but an awesome puddle or splotch of melted babbitt. The other extreme, a thrust bearing not taken to complete failure, often shows (Fig. 10-4) either some darkened areas at the minimum clearance point, spalling, flakings, etc., or the bearing looks to be in original as-received conditions.

CONCLUSION

It is hoped that this review of normal thrust setting procedures, thrust or rotor movement monitoring, plus some of the practices and beliefs in thrust failure prevention, will be helpful to those establishing a failure prevention system in this area.

LITERATURE CITED

[1] Jackson, C., "An experience with thrust bearing failures," *Hydrocarbon Processing*, January 1970, pp. 107-110.

Balance Rotors
by Orbit Analysis

It is necessary to balance some equipment in the field, in its own bearings, and at operating speeds. These techniques often can be used in the shop as well. Orbit balancing is one example. Balancing turbomachinery, as mounted, at least gets the rotor away from journal marking anti-friction bearings, although convenience is hampered. However, many machines—blowers, turbomachinery with external trim balance rings, and pumps—can be balanced in place.

Using portable field balancing equipment, an original vector (O) is plotted from the displacement (mils peak-to-peak) and the phase angle measured (Fig. 11-1). A trial weight is added. For example, the trial weight can be based on approximately ten percent of the journal static weight as a force value.

$$T_W = \frac{0.1 \text{ (Journal weight, lbs.)}}{1.77 \text{ (Balancing speed, rpm/1,000)}^2 (R)}$$

T_W = Trial weight, ounces
R = Radius of correction, inches

If the phase lag of the analyzer or the machine is known plus the rotor sensitivity (ounce-inches per mill correction), this first trial weight can easily be installed very near the light spot on the rotor (180 degrees opposite the heavy spot).

The pencil method of balancing. Many craftsmen and engineers have tried to balance a machine using a pencil to mark the shaft. Given a slow speed, large deflections, a steady hand and some nerve, this could have been successful (see Fig. 11-2). A weight is simply added opposite the point of shaft deflection or pencil mark on the shaft.

This article will present an accurate electronic approach to the pencil method of balancing. Electronically, an orbit path of the shaft can be generated. Also, electro-mechanically, a phase reference mark can be produced to show where the shaft is at a particular time. The author's experience is confined to machines with hydrodynamic bearings.

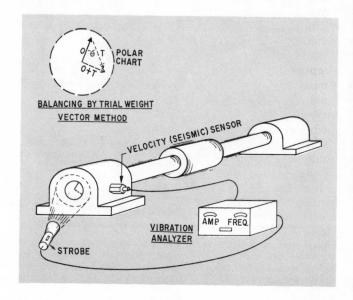

Fig. 11-1—Portable field balancing equipment can be used.

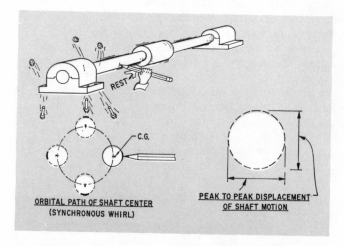

Fig. 11-2—You can balance with a pencil if you have slow speeds, large deflections, and a steady hand.

Electronic balancing. Displacement sensors will be needed to measure shaft motion. The sensors often provided on most modern, critical equipment to measure vibration can be successfully used, with a little adaptation, to provide a good balance. Further, the phase angle to apply the first weight correction can be determined with a reasonable degree of accuracy (plus or minus 5 degrees). Applying the first weight correction should always be done with great care so that imbalance is not drastically increased. In addition, using a dual-beam oscilloscope, pairs of vibration probes, and a phase-reference probe and mark will give a very good indication of the actual criticals for flexible rotors. Also, the extent of a second or higher critical *skirt* can be easily observed.

The technique presented has been used on one laboratory model, another larger multi-disk rotor model (approximately 4 feet long, and a 70-inch blower wheel unit rated at 450 hp. With some 70 channels of vibration being installed on new equipment and each train having a *key-phase* reference on the driver, it is expected that field balancing will be used. In addition, problem areas will be better defined.

Preparation. Two inductive displacement sensors are mounted at each bearing, located 90 degrees apart, at the same radius. A shaft discontinuity (notch, hole, pin, keyway) is located or created. Another sensor probe is mounted radially on the stator in a transverse plane to the shaft axle passing through the path of this discontinuity, i.e., the probe looks at the discontinuity as it passes by once per revolution of the shaft.[1]

If each of the two-bearing mounted sensor-probes can be mounted on the true vertical and horizontal position, it will greatly simplify the orientation of the human mind. However, if they cannot, which is typical for axial-split equipment, a location near the horizontal and vertical is fine (Fig. 11-3). They should be 90 degrees apart. With vertical shafts, one can reference to anything convenient.

The sensors are calibrated for mils (0.001 in. = 1 mil) peak-to-peak displacement versus a DC voltage output. We have used 200-mv negative DC peak-to-peak (p-p) voltage per mil of peak-to-peak displacement. This transfer ratio (input signal to output signal) allows the output voltage to be connected to a 1-meg or greater source impedance, dual-beam oscilloscope. The horizontal sensor is connected to the positive polarity input of the horizontal amplifier on the scope. The scope precalibrated voltage sensitivity is set at 200 mv/cm. The vertical sensor output is connected to the vertical amplifier jack (+). The phase reference probe (hereafter referred to as the *key-phase probe*) is oriented over a notch on the shaft (Fig. 11-4). The output signal of this sensor is connected to the trigger jack of the scope for synchronizing the scope, but more importantly, it is connected to the Z-axis input of the Cathode Ray Tube (CRT). This connection must be AC coupled to the Z-input, and the normal ground shorting bar must be removed. If the scope is AC coupled at this point (typical of Tektronix), there is no problem. If it is DC coupled (typical of Hewlett-Packard), a capacitor is needed in series with the signal. Shielded coaxial cables

Fig. 11-3—Typical probe mountings at bearings.

Fig. 11-4—The key-phase probe acts as a reference when the notch on the shaft passes the probe.

should be used, and only *one* earth ground is to be used for all equipment. You will note in the schematic (Fig. 11-5) that an amplifier is used to give approximately 10 to 12-v output on the key-phase circuitry. This *spike* in voltage will assure that the trace on the CRT is *blanked* when the notch passes the sensor probe, thus causing an amplified voltage change. Blanking is injecting an intensity change to the trace on the CRT to mark a position. A boost in negative voltage is given as the notch enters the probe face area (increase in gap) and a less negative (positive slope) burst of voltage occurs when the notch's trailing wall passes the probe face (decreases in gap). This causes *bright* and *break* spots (Fig. 11-6). The order will depend on oscilloscope internal circuitry.

The orbit. The Lissajous, or orbit pattern, displayed on the CRT of the scope represents the actual motion of the shaft within the bearing. Four typical orbits are shown in Fig. 11-7.

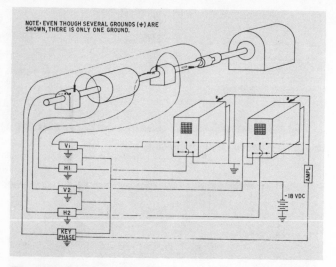

NOTE: EVEN THOUGH SEVERAL GROUNDS (⏚) ARE SHOWN, THERE IS ONLY ONE GROUND.

Fig. 11-5—Typical two-bearing machine with a driver as used in measurements.

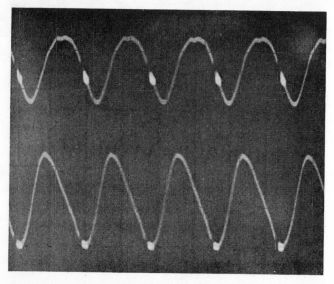

Fig. 11-6—Key-phase mark 'blanks' the time sweep of vertical and horizontal shaft motion.

The first orbit represents a typical motion of a shaft within a bearing. The amplitudes in the horizontal and vertical direction are even and the total peak-to-peak displacement is 1½ mils in the horizontal and vertical plane with no indicated higher multiple frequencies of the running speed.

In the second illustration, the circular motion is now elliptical and predominately in the horizontal plane, indicating that the motion of the shaft is two-to-one in the horizontal plane versus the vertical plane, i.e., 2 mils peak-to-peak horizontally and 1 mil peak-to-peak vertically.

In the third illustration, the motion of the shaft is from 11 o'clock to 5 o'clock with a component of double frequency indicated by the small loop in the diagram. This particular diagram was recently recorded from a

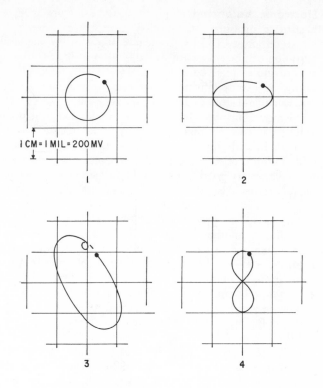

1 CM = 1 MIL = 200 MV

Fig. 11-7—Typical orbit patterns.

10,000-hp steam turbine in which the bottom-half bearing did not match the upper-half bearing. This committed the shaft in one orientation and forced a function twice the rotating journal speed.

The fourth illustration indicates a predominately vertical shaft motion but with a very strong component at twice the fundamental running speed frequency, hence the figure eight pattern.

For orientation, one should review the phase relationship of Lissajous patterns.[3] A quick review is shown in Fig. 11-8. The 90-degree lagging signal fixes the CRT phase shift at 90 degrees which is the plane of the CRT phosphor-coated surface.

Fig. 11-9 shows the three signals in voltage (ordinate) versus time (abscissa) as taken from a dual-beam oscilloscope. The top trace shows the key-phase mark as the notch goes by the probe. The middle trace is the vertical-output. The lower trace is the horizontal-probe output. (Calibration of the middle and lower trace is 1 cm [large division] = 1 mil.)

Fig. 11-10 shows the orbit pattern, without the key-phase mark, blanking the x-y pattern on the CRT. Fig. 11-11 shows the same pattern with the blanking mark.

Care should be taken to establish the direction of the orbit with relation to one's position viewing the machine to be measured. We always view, from outboard, the driving end facing the driven machine. A *forward* orbit would then be one that rotates in the same direction as the shaft.

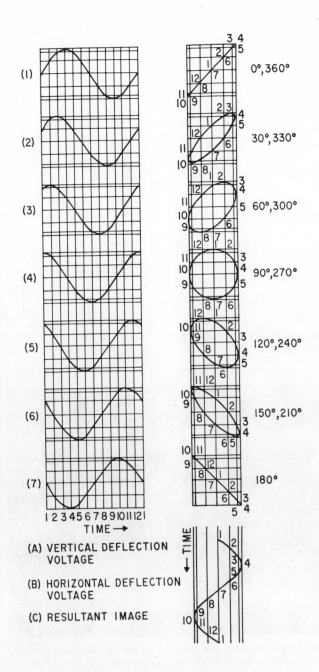

(1)

(2)   0°, 360°

(3)   30°, 330°

(4)   60°, 300°

(5)   90°, 270°

(6)   120°, 240°

(7)   150°, 210°

           180°

1 2 3 4 5 6 7 8 9 10 11 12
TIME ⟶

(A) VERTICAL DEFLECTION
      VOLTAGE

(B) HORIZONTAL DEFLECTION
      VOLTAGE

(C) RESULTANT IMAGE

Fig. 11-8—Lissajous phase relations.

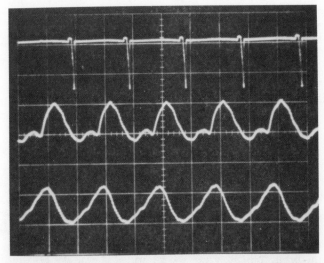

Fig. 11-9—CRT display showing the key phase (spike), the vertical and horizontal time traces.

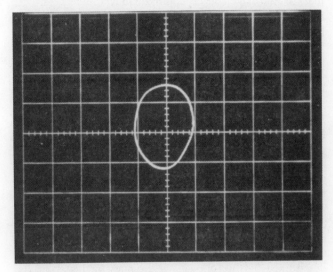

Fig.11-10—Orbit without a key-phase mark.

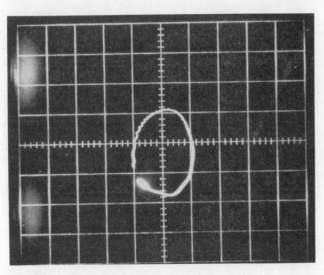

Fig. 11-11—Orbit with key-phase mark blanking.

Fundamentals to balance. To recap, the orbit represents a graphical picture of the shaft motion pattern. The key-phase mark represents *where* the shaft is at the very instant the notch passes the probe (sensor). I will remind those familiar with stroboscopy that there are no phase lags in this circuit as with filters and other RC&RL time-lag contributors. The *bright* and *break* marks occur as the gap changes through the *notch start* and *notch end*. This helps check the orientation of the orbit and also gives you a measure of the notch length (circumferential).

Second Intuitive Rule. Below the first critical, the mark on the orbit represents the location of the heavy spot of the shaft relative to that bearing. This point is difficult to see, yet simple once it is understood; i.e., the shaft must be wherever it is because of either external forces or mass imbalance. Limiting this discussion to imbalance, the shaft is displaced by imbalance. The mark shows *where the shaft* is at that precise instant when the notch passes the probe.

Therefore, if one would stop the machine and turn the shaft until the notch lines up with the probe, the angular position of the shaft is satisfied. Then, laying off the angle from the pattern taken on the CRT gives the *heavy* spot for correction. Weight can either be subtracted at this point or added at a point 180 degrees diametrically opposite, on the shaft. The customary guidelines[6] for weight-radius amount can be used. The orbit diameter will reduce as correction is applied. Should too much weight change be given, the mark will shift across the orbit, indicating the weight added is now the greatest imbalance.

Above the first critical, the rules change. The key-phase mark would have shifted approximately 180 degrees when the shaft mode of motion changed from translational to pivotal. Therefore, the phase mark will appear opposite the actual heavy spot and the weight addition would be on the key-phase mark position. *Note*: If probes are located external to shaft nodal points (generally several inches external to bearings), the preceding rules reverse since the shaft has reversed its position.

Balancing. To balance a rotor, bring the rotor to a desired balance speed. This speed should not be near a critical, as the phase shifting from the rotor modes should not confuse the motion due to imbalance. Couple correction, e.g., oz.-in., is made according to the rules explained. The machine is rerun at the speed to either confirm the correction or to make another adjustment for better balance tolerance.

The correct angle of the key-phase mark must be determined by the following techniques for better orbits:

1. Lay a major diameter axis at the greatest beam of the orbit.
2. Circumscribe a circle about the diameter in Step 1.
3. Through the key-phase mark on the orbital trace, draw a perpendicular line to the major axis of Step 1.
4. Extend this line radially from the major axis, if needed, until it bisects the circle drawn in Step 2.
5. The angle taken either clockwise or counterclockwise from the vertical, as an example, is to be marked on the rotor when the machine is stopped and the notch orientated to the key-phase probe as earlier described. (Care must again be given to hold the original orientation, e.g., facing rotor from driver end).

Fig. 11-12 outlines this procedure. Fig. 11-13 illustrates the angle being marked on the rotor after orientation. Either photographing the orbit on the graticule or the use of a storage scope greatly assists this operation. It takes a few minutes. A permanent record is also provided.

It may be convenient to plot an orbit from the horizontal and the vertical traces versus time sweeps; i.e., x-t

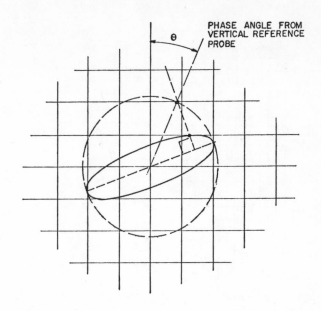

Fig. 11-12—Phase angle measurement of key phase from vertical axis.

Fig. 11-13—Marking the phase position on the shaft with the notch in line with the key-phase probe—shaft at rest.

and y-t. A shaft affected by many forcing frequencies or an electrical system with significant noise may cause an orbit not to be clear. A further development of this technique would include a phase-angle meter to provide a digital read-out of the phase. The scope may still be necessary to determine the quadrant. Varying speed has little effect on balancing which is a real advantage when working on induction motor driven equipment which *slips* or turbine drives which *hunt* in operation.

Two tunable band-pass filters can be used, with one filter in each circuit, to filter out frequencies other than that of the shaft's synchronous frequency. These filters give a very clear and unaltered phase angle; however, each of the two filters *must be accurately tuned to peak resonance* at the shaft's fundamental rotating frequency.

Illustrated procedure. The rotor model shown in Fig. 11-14 will be used in a demonstration of orbit balancing. When the tape on the rotor is in the top position (Fig. 11-14) it signifies that the key-phase notch (Fig. 11-4) is straight up, at the 12 o'clock position, in line with the key-phase probe. The scope will show a dot when the shaft is not rotating.

To illustrate the procedure, a balanced shaft will be used as a beginning. A known weight will be added to the rotor to unbalance it; then corrective weights will be subtracted to rebalance the rotor. Finally, a weight will be added at 180 degrees out of position to show that the key-phase reference will shift 180 degrees.

Fig. 11-14 shows two weights of 0.17 ounces each added at 1.75 inches radius (0.6 oz.-in.) and at the 9 o'clock position. On Figs. 11-15 and 11-16 you will note that at

3,050 rpm there is a 2.8 mils vertical and 3.0 mils horizontal motion with the key-phase mark at 9 o'clock.

Next, one weight is removed from the 9 o'clock position (approximately 0.3 ounce-inches), see Fig. 11-17. Figs. 11-18 and 11-19 now show the motion to be 1.8 mils vertical and 1.2 mils horizontal with the phase mark still at 9 o'clock.

The last weight is removed with the vibration reduced to 0.5 mils vertical and 0.4 mils horizontal (Figs. 11-20 and 11-21). Actually, the phase is shifting to about 5-6 o'clock where the greatest bearing clearance exists.

Finally, one weight is reapplied at the 3 o'clock position, Fig. 11-22. The heavy spot now changed to 3 o'clock and the key-phase mark changed respectively (see Fig. 11-23). The latter photograph was taken as an afterthought on another scope for the non-believers. Note that the phase responded the same, however the *bright* spot now precedes the *break* in the counterclockwise position as discussed earlier.

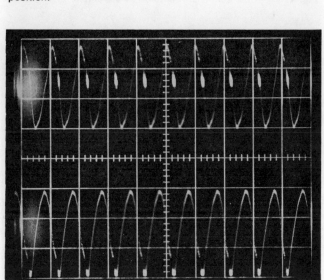

Fig. 11-14—Two known weights are added at the 9 o'clock position.

Fig. 11-15—Top trace shows 2.8 mils vertical motion and bottom shows 3.0 mils horizontal motion.

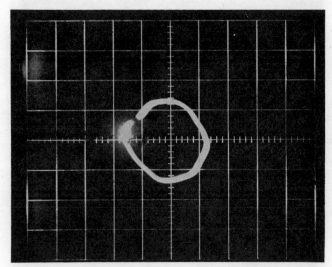

Fig. 11-16—Orbit pattern of Fig. 15 shows the same motions plus the key-phase mark at 9 o'clock.

Fig. 11-17—One weight is removed from the 9 o'clock position.

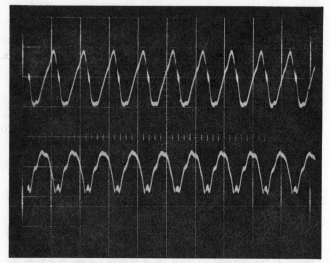

Fig. 11-18—The top trace shows the vertical vibration reduced to 1.8 mils and the bottom trace shows the horizontal vibration reduced to 1.2 mils.

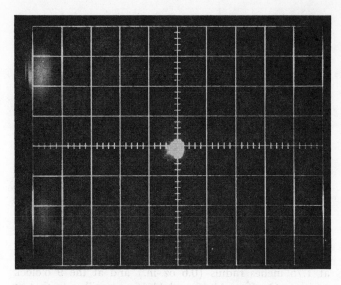

Fig. 11-21—The orbit pattern of Fig. 20, now almost a dot, shows the same low imbalance.

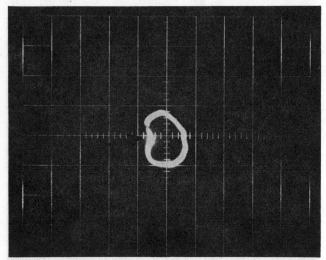

Fig. 11-19—Orbit pattern of the Fig. 18 shows the same vertical and horizontal motions.

Fig. 11-22—One weight is reapplied at 3 o'clock.

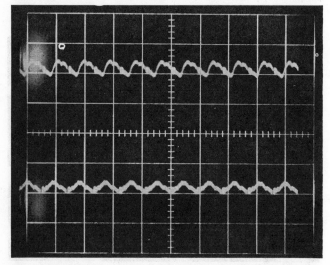

Fig. 11-20—With the last weight removed, the vertical vibration is 0.5 mils and the horizontal is 0.4 mils.

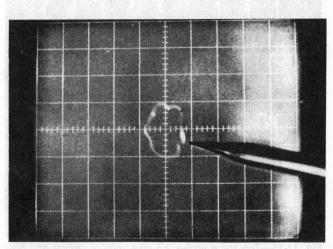

Fig. 11-23—The orbit pattern from Fig. 22 shows the key-phase mark at 3 o'clock.

Other orbit information. The orbit offers many other bits of information on the rotor response: oil whirl, rubs, multiple frequencies, base resonance [4] ($\frac{1}{4}$, $\frac{1}{3}$, $\frac{1}{2}$ fundamentals), alignment double frequencies, and critical skirts to name a few. A digital tachometer reading the key-phase pulses is a convenient accessory. It is not the intent here to go into this area but the author would encourage more practical work in this field.

Oil Whip Correction. One of the recent uses of orbit analysis is in the proper location of the hydrodynamic oil-film anti-oil-whip dam to prevent severe oil whip. One of the criteria for location of the scraper-type dam is to place it in the line of action of the shaft. The orbit easily defines the line of action of shaft motion. For example, in Fig. 11-7 (3rd) the proper action of the oil dam would be approximately 30 degrees from the vertical centerline of the bearing.

The orbit approach to balance is valid. It will work. It is simple to use and understand, but it needs more refinement. Proper installation of vibration probes can provide the major tooling required with the only addition being a key-phase probe and the additional requirement of a dual-beam oscilloscope and camera at less than $1,500.

ACKNOWLEDGMENT

Based on the paper 70-Pet-30 originally presented to ASME in Denver, Co., Sept. 14, 1970.

LITERATURE CITED

[1] Bently, D. E., Orbits, Bently-Nevada Corporation, Minden, Nev., 1969.

[2] Sternlight, B., and Lewis, P., Vibration Problems With High Speed Turbomachinery, ASME Paper 67-DE-8.

[3] Malmstadt, Enke, and Toren, Electronics for Scientists, W. A. Benjamin, Inc., 1963; Plus Radar Electronic Fundamentals, U. S. Government Printing Office, Washington, D. C., 1944.

[4] Hartog, D., Mechanical Vibrations, McGraw-Hill.

[5] Harris and Crede, Shock and Vibration Handbook, Vol. 3, McGraw-Hill.

[6] Muster, D., and Flores, B., Balancing Criteria and Their Relationship to Current American Practice, ASME Paper No. 69-Vibr.-60.

[7] Jackson, C., New Look at Vibration Measurement, Hydrocarbon Processing, Vol. 48, No. 1, Jan. 1969.

[8] "IRD Instruction Manual," IRD, Inc., Columbus, Ohio.

Optimize Your
Vibration Analysis Procedures

A formal vibration analysis system can save millions in rotating machinery failures. The type, speed, and critical nature of the equipment should dictate the vibration program needed. There are many spin-offs from such a system, e.g. balance equipment and procedures, alignment equipment, bearing gauging and maintenance procedures. The program outlined is for a plant employing seventeen hundred people with 120,000 plus horsepower in compression equipment, 1600 pumps and other special equipment. Fourteen thousand horsepower is the largest single component size, with 100,000 horsepower in turbomachinery at speeds up to 52,000 RPM. This chapter covers the background, present system, practices, and some recent experiences of a system in existence for 10 years. Five engineers and one engineering aide (average engineer experience is twenty years) make up our analysis group.

Prior to the use of formal vibration analysis techniques, a vibrating light vibrometer, a nickel, or sensitive hands were our main analyzers.

The first instrument purchased was a hand-held vibrograph. It gave us a pressure sensitive tape of the vibration waveform for records and frequency displacement analysis. It was totally mechanical, small, lightweight and could be used on housings, structures, or directly on shafts with a "fish tailed" shaft stick. The tape can easily be copied for reports. This instrument is still in continued use. It is most useful in confirmation of non-periodic malfunctions such as oil whirl.

Our first electronic vibration and balancing analyzer used a seismic, or velocity, transducer and cost about $2800. This unit put us into the preventive maintenance business on a large scale. I might add that that was not the original intention. As a Mechanical Development Group under Engineering Services, we intended to simply define problems by proper measurements. A program of 90 or 180 day vibration surveys was established for critical rotating machinery. The scheduling was complicated by an equal amount of emergency type inspections. Historical data files were established. These files contained pure and manually tuned vibration spectra and unfiltered velocity

level trend charts. These trends were recorded at three positions in the shaft bearing area, horizontal and vertical in the radial direction, and in the axial direction (parallel to the shaft).

The trend charts are helpful because they illustrate how far gradual excursions are drifting over a given period of time. The chart shows the diminishing health of a machine. The pure (unfiltered) velocity data is the most reasonable plot to make. If one will recall simple harmonic motion, peak velocity is the only product term at the first power of frequency and peak amplitude.

$$\text{Displacement} = X = A \,(\sin wt) \quad = A \quad \text{Peak Value}$$

$$\text{Velocity} = \dot{X} = Aw \,(\cos wt) \quad = Aw \quad \text{Peak Value}$$

$$\text{Acceleration} = \ddot{X} = Aw^2 \,(\sin wt) = -w^2A \quad \text{Peak Value}$$

This type of analyzer evolved into the current lightweight solid state analyzers with frequency meter, displacement, velocity, acceleration and sound measure, strobe light tunable filter for 10-1,000 Hz use.

Proximity measure. As long as the equipment was accessible and the force transmission from rotor to bearing housing was high, this program worked well. However, a new generation of machines were accompanying large single train plants.

- Normal speeds were now 10,000-52,000 RPM.

- Ratios of total machine weight to rotor weight were running 20:1 up to 125:1.

- Coupling and shafting were total enclosed for continuously lubricated coupling drives.

- Equipment was operating generally above the first critical speed and often above the second critical.

- Horsepower per train was typically 10,000 to 30,000 horsepower. Other companies are up to 60,000 horsepower per train.

Protection monitoring systems using eddy current (inductive type) sensor probes became necessary. We have approximately 200 of these probes installed on bearing housings observing shaft motion relative to the bearing. These "electronic eyeballs" give good sensitivity for early detection of machine malfunctions and are perfect for vibration analysis.

Protection monitoring is all together a different philosophy from vibration analysis though many view it the same. However, if one will provide a well designed monitoring system, the data can be easily obtained for vibration analysis.

Present analysis system. About 1966, we adopted a program called MACE (Measurement, Analysis, Correction, Engineering). This was viewed similar to the fire triangle of fuel, oxygen, and ignition, i.e. remove one of the legs and you do not have a fire. Removing one component of MACE renders it incomplete.

We attempted to expand our measuring hardware at a rate equal to our understanding. As a result, we are adding capital to our department at approximately $10,000 per year. In 1966, we had about $10,000 in portable measuring equipment, about six plant installed seismic type protection monitors, and a small lab building. In 1973, we had approximately $85,000 in portable equipment. In 1978, we have about $167,000 in portable diagnostic equipment as indicated by the following list:

- Two real time spectrum analyzers to include: dual channel comparisons, orbits, 400 line displays, peak hold, transient capture, scale comparison — $ 20,000

- Two eight-channel FM recorders and preamplifiers — $ 22,000

- Two digital vector tracking filters with dual channel tuned orbit capabilities — $ 16,000

- Two battery powered four-channel direct-FM cassette tape recorders for 10 Khc and 1 Khz frequency response ranges — $ 7,200

- Two- and four-channel AM tape recorders — $ 1,000

- Five manual tunable vibration analyzers — $ 10,200

- Vibrometer, vibrographs — $ 1,000

- One 1/3, 1/10 octave band vibration/sound analyzer with recorder — $ 3,100

- Alignment—33 channels proximity — $ 15,000

- Alignment—3 optical instruments and accessories — $ 8,000

- One eight-channel strip recorder with max. speed of 200 mm/sec. This instrument is essential in studying time related multiple variations in measurements — $ 11,000

- Two lab oscilloscopes and accessories — $ 5,000

- Strain and fatigue gauge facilities — $ 2,000

- Test instruments and cables — $ 3,000

- Sensors—piezoelectric, geophone, inductive — $ 5,000

- Three digital tachometers — $ 2,000

- X-YY' recorders (amplitude vs phase vs RPM plotting) — $ 6,000

- Two 35mm cameras and lens plus three Polaroid cameras (two are scope cameras) — $ 2,400

- Long wheel base van with work table — $ 7,000

$167,000

One might question the amount of investment, but our rotating equipment maintenance cost is $2 MM/year. The compressor repair costs are only $.6 MM. The turbo compressor maintenance cost with drivers is approximately $3.00/horsepower/year. The 30,000 horsepower in an eight-year-old, 1,000 ton/day methanol plant plus 20,000 hp in a two year acrylonitrile plant operate less than $2.00/horsepower/year.

Practices. The API Standards recently published are good standards for rotating equipment. Monsanto contributes to these standards and uses these standards as do the refineries. Some of the practices for protection monitoring are:

- Two 90° circumferential oriented probes are attached at each bearing of critical equipment.

- A phase marking, one event per revolution, probe is provided on each driver (second driver on gear units).

- Dual channel "high read" alarm and trip readouts are provided.

- Should all measuring points not be monitored, the information is routed to the control room for diagnostic measure.

- Explosion proof, NEMA 7 type enclosures are limited to machinery location. NEMA 4 are often used.

- An attempt is made to provide all installations, allowing probe replacements while operating.

- Leads are protected in rigid and flexible conduit.

- Thrust displacement probes generally read the thrust collar or the shaft end near the thrust collar.

- Thrust displacement probes are always on automatic shutdown with dual redundant logic, i.e. both probes must vote to trip. Calibration of 200 mv/mil (8 v/mm) or 100 mv/mil (4 v/mm) is provided with a minimum linear range of 80 units (2.032 mm) specified (API 670). The meter scale is 1 mm-0-1 mm (2 mm total).

- Five mils 125 μm (peak/peak) radial vibration meter range is normally specified.

- Two level alarms are always specified. The first warns and/or allows call-out for troubleshooting. The second level commits either automatic or manual shutdown.

- Electrically deenergized systems which are energized to trip are preferred over continuously energized systems.

- Coaxial connectors are kept to a minimum and these are wrapped in Teflon tape and incapsulated in RTV silicone rubber.

- Sensor systems are standardized, generally with 5 meter probe lead lengths. Probe lengths are one meter.

- The monitoring hardware is located on the same side of drivers and driven equipment.

- Vibration proximity sensors are arranged to be 45° from the horizontal joint and in the upper half, i.e. @ the 10:30 and 1:30 o'clock position.

- The horizontal sensor must be 90° to the right of the vertical sensor for proper analysis without special conditioning. The left 10:30 o'clock sensor will always be the vertical and the right 1:30 o'clock sensor will be declared the horizontal.

- Great care is given to safety by using grounded system with care that only one ground exists, i.e. no ground loop noise.

There is a philosophy on vibration analysis which must be realized. It needs to grow in normal steps. One can always measure more than can be explained. The training and supporting education comes slowly. Many times we investigate a pin pointed area not knowing what we will find, but having at least pin pointed the area. At some point, science finally overtakes art, but there is a lot of art left. Intuition becomes a strong tool. Often, we can anticipate what will happen without fully understanding why. To be effective, one must be thoroughly familiar with machine internal construction.

EXPERIENCE

One might presume that many "saves" must occur to justify the continued purchase of expensive portable and fixed vibration equipment. This is true. Some selected results will be listed to show the type of coverage rather than the amount.

Case I. During commissioning, a 9,000 horsepower, 11,000 RPM steam turbine developed a 44% non-synchronous oil whirl instability that eventually reached approximately 5 mils peak/peak vibration, journal to bearing. Several tests and orbit studies indicated the instability was initiated by steam valve sequence loading. The shaft first orbits elliptically from 11 to 5 o'clock. Then after the next quadrant of valves opens, the orbit shifts to 9 to 3 o'clock.

The pressure dam type bearing locking pin was relocated to damp the 9 to 3 o'clock motion and the machine was operated here for two weeks until a 4 lobe bearing design by the turbine builder was manufactured. This bearing was installed in one shift with no loss in production.

Case II. A compressor rotor on a 14,000 horsepower train was slowly fouled with a clay like deposit over six months, suddenly unbalancing the rotor during slough off.

The synchronous response jumped from 0.6 mils peak-to-peak to 2.7 mils peak-to-peak. The equipment was closely monitored for two more months for a scheduled shutdown and the spare rotor installed. During the next six months, the process separation design equipment was revised and a second rotor switch with separation revision put us back in control. Unbalance amounts confirmed rotor response tests during the mechanical run test at the factory before shipment to job site.

Case III. A 5,000 horsepower motor-gear-compressor train was shutdown fifteen seconds after commissioning start. A severe rub was detected by the vibration probes, limiting the damage to the compressor sleeve and labyrinth seals (Fig. 12-1). The unit was repaired in forty-eight hours. The reason for the rub was a mismatched top seal half.

Case IV. Severe misalignment of an 8,000 horsepower turbine-compressor train caused a coupling-end unbalance which rose to 4 mils peak-to-peak. Shutdown and inspection at the coupling hub-shaft keyway section indicated a fatigue break 60% through the shaft (Fig. 12-2).

Case V. Startup of a refrigeration compressor accompanied by vibration and hot alignment measures indicated that the compressor discharge pipe to the condenser, moved the compressor over 100 mils horizontally out of alignment with the drive turbine. The refrigeration design manufacturer rechecked his expansion joint calculations and found an error in number of convolutions required. Vibration signals were typical for misalignment, i.e. high axial vibration and twice running speed radial vibration.

Case VI. A turbine on a five case train developed 3 mils peak-to-peak vibration at 10,300 RPM. The coupling

Fig. 12-1—Labyrinth seal rub on a centrifugal compressor. Machine shutdown 14 seconds after startup.

Fig. 12-2—Rotor shaft crack at coupling keyway on 9,000 horsepower steam turbine.

shaft keyway was short a full key, thereby causing an unbalance condition.

Case VII. A 14,000 horsepower, 11,000 RPM steam turbine was field balanced with two external balance rings to reduce synchronous vibration levels from 2.6 mils peak-to-peak to 1.25 mils peak-to-peak in one shift without disassembly. This extended this machine's two-year run. Seismic analyzer with strobe reference were coupled with orbital phase marking to place weight and correct in a minimum of two runs. A two plane couple correction was made one month later in less than one hour to effect a 1.1 mil peak-to-peak vibration level.

Case VIII. Due to an improperly positioned discharge-end bearing housing, the thrust on a barrel compressor had worn the balance piston labyrinth clearance to triple original clearance. The thrust deflection gradually progressed until at 17 mils from original load point, the compressor was shutdown with a smeared but partially failed bearing but undamaged compressor rotor assembly. Static balance chamber pressure confirmed loading but thrust bearing exit oil temperatures did not.

Case IX. At another location, a 15,000 horsepower steam turbine was shut down manually on both thrust alarms plus the radial vibration alarms. The turbine had a heavy deposit of salt on two stages (Fig. 12-3). The thrust bearing had failed and the blading shrouds contacted the diaphragms cutting into them about ⅛″. Figure 12-4 shows damage to the thrust collar from overheating. Though the trailing edges on two blading rows were drawn and required grinding, the rotor was saved and put back in service one week later.

The company has paid from $⅓ MM to $1 MM on this type of wreck in the past.

Case X. At another location, the coupling housing filled with oil due to an improperly sized gasket which choked the drain. This caused an overfilling which took several hours from startup. A one-half running frequency vibration was the indication. The machine was shut down to inspect the bearing. A couple of energetic maintenance men started lifting the top half of the coupling guard only to find the flooded condition. It was a lucky accident. The bearing was suspect but the coupling, for the first time in its life, saw the opportunity to become a bearing, though a poor one.

Fig. 12-3—Heavy salt deposits on steam turbine rotor blading which caused thrust bearing failure.

Fig. 12-4—Severe thrust collar damage resulting from salt deposits on steam turbine blading. Note the heat tears on the thrust collar face.

This brings up an interesting thought about vibration analysis. Several malfunctions may send the same type of frequency signal, i.e. one might think of this as one or more radio stations transmitting on someone else's frequency.

A loose thrust collar, loose coupling hub, double kinked shaft, or misalignment might well send a twice running speed frequency signal.

Because 60% of our vibration problems resulted from misalignment over our first five years of vibration analysis, we began a precise measurement of shaft and casing alignment using the eddy current probes. We have 33 channels

of this equipment, plus 3 optical instruments to aid in this work. Both techniques have been published.[1,2]

Severity. The most difficult problem in vibration analysis is establishment of severity limits. Machines, much like people, have different sensitivities and resistances to vibration forces.

In general, we have found that 0.2 inches/second (peak) velocity to require alarm action on bearing housing seismic type reading. Greater values, e.g. 0.3 to 0.5 ips will normally force shutdown. The shaft-to-bearing relative measure limits are different. With clean shafts, we generally set high speed machines with 2.2 mils peak-to-peak warning and 4.2 mils peak-to-peak shutdown limits.

Rotor thrust displacement monitors are set with 15 mils over normal load deflection as an alarm and either 20 or 25 mils past normal load deflection as an automatic shutdown position. Based on plant experience, it is felt that you cannot risk manual shutdown in thrust. The normal failure time is thirty seconds!

ACKNOWLEDGMENT

The author wishes to acknowledge the contributions of Engineers E. J. Opersteny, Paul S. Gupton, Carl A. Duhon, Howard Blackburn, and James H. Ingram, Materials & Mechanical Technology, Monsanto, Texas City, Texas to the MACE program as reported. Copyrighted by and originally presented at the Institute of Electrical and Electronic Engineers, Petroleum Division Conference, September 1973, Houston, Texas.

LITERATURE CITED

[1] Jackson, C., "How to align barrel-type centrifugal compressors," *Hydrocarbon Processing*, 50, No. 9, p. 189, September 1971.

[2] Jackson, C., "Successful shaft hot alignment," *Hydrocarbon Processing*, 48, No. 1, p. 100, January 1969.

Appendix 1

Vibration Unit
Conversion Nomograph

This chart allows one to correlate peak-to-peak displacement (inches) vs peak-to-peak velocity-ipsp vs peak acceleration in "g" units vs a known frequency in hertz (cps).

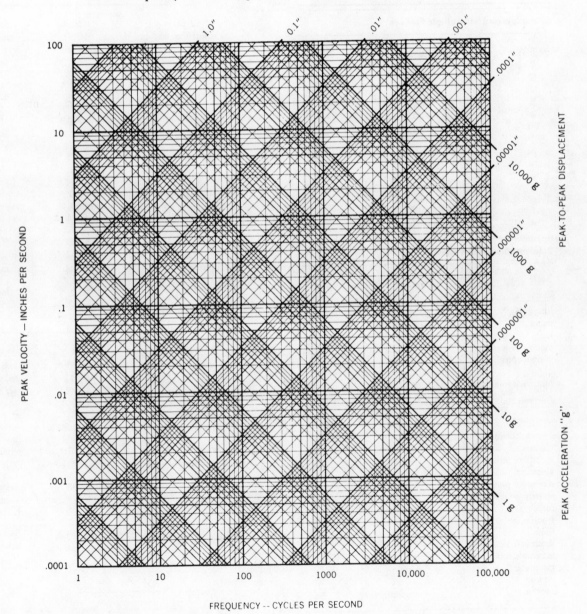

Courtesy of Dytronics Company, Columbus, Ohio.

83

Appendix 2

Conversion Factors*

A. Conversion Factors

1. Multiple and Submultiple Prefixes

Symbol	Prefix	Multiple
T	tera	10^{12}
G	giga	10^9
M	mega	10^6
k	kilo	10^3
h	hecto	10^2
da	deca	10
d	deci	10^{-1}
c	centi	10^{-2}
m	milli	10^{-3}
μ	micro	10^{-6}
n	nano	10^{-9}
p	pico	10^{-12}
f	femto	10^{-15}
a	atto	10^{-18}

2. International (SI) System of Units

Basic Units

Quantity	Name	Symbol
length	meter	m
mass	kilogram	kg
time	second	t
electrical current	ampere	A
temperature	kelvin	K
	Celsius	°C
luminous intensity	candela	cd

Derived Units (with special names)

Quantity	Name	Symbol
electric charge	coulomb	C
electric capacitance	farad	F
electrical inductance	henry	H
electrical potential	volt	V
electrical resistance	ohm	Ω
energy (work and heat)	joule	J
force	newton	N
frequency	hertz	Hz
illumination	lux	lx
luminus flux	lumen	lm
magnetic flux	weber	Wb
magnetic flux density	tesla	T
power	watt	W
pressure	pascal	Pa

3. Length

Multiply	by	to obtain
angstrom (Å)	10^{-10}*	meters
feet	0.304 80 *	meters
	12.0 *	inches
inches	1 000	mil
	25.40 *	millimeters
	0.025 40 *	meters
	.083 33	feet
kilometers	3 280.8	feet
	0.621 4	miles
	1 094	yards
meters	39.370 *	inches
	3.280 8	feet
	1.094	yards
miles (statute)	5 280 *	feet
	1.609 3	kilometers
	.8694	miles (nautical)
millimeters	0.039 37 *	inches
miles (nautical)	6 076	feet
	1.852 *	kilometers
yards	0.9144 *	meters
	3.0 *	feet
	36.0 *	inches

4. Area

Multiply	by	to obtain
acres	43 560 *	sq. feet
	4 047	sq. meters
	4 840	sq. yards
square cm	0.155 0	sq. inches
square feet	144.0 *	sq. inches
	0.092 90	sq. meters
	0.111 1	sq. yards
square inches	645.16 *	sq. millimeters
square kilometers	0.386 1	sq. miles
square meters	10.764	sq. feet
	1.196	sq. yards
square miles	640 *	acres
	2.590	sq. kilometers

*Exact

*From *Endevco Dynamic Handbook* by Endevco Dynamic Instrument Division, Pasadena, California.

5. Volume

Multiply	by	to obtain
acre-foot	1 233.5	cubic meters
cubic cm	0.061 02	cubic inches
cubic feet	1 728 *	cubic inches
	7.480	gallons (U.S.)
	0.028 32	cubic meters
	0.037 04	cubic yards
cubic inches	16.387	cubic cm
	0.017 32	quarts (liquid)
cubic meters	35.315	cubic feet
	264.2	gallons (U.S.)
	1.308	cubic yards
	1 000	liters
cubic yards	27.0 *	cubic feet
	0.764 6	cubic meters
gallons (Imperial) (U.K.)	277.4	cubic inches
	1.201	gallons (U.S.)
	4.546	liters
gallons (U.S.) (liquid)	231. *	cubic inches
	3.785 4	liters
	4.0 *	quarts
quart (U.S.) (liquid)	0.946 3	liters

6. Mass

Multiply	by	to obtain
carat	0.200 *	grams
grams	0.035 27	ounces (avdp.)
kilograms	2.204 6	pounds (avdp.)
ounces (avdp.)	28.350	grams
pound (avdp.)	16.0 *	ounces (avdp.)
	453.6	grams
stone (U.K.)	6.350	kilograms
	14	pounds (avdp.)
ton (long) (2240 lb)	1 016.0	kilograms
ton (short) (2000 lb)	907.2	kilograms
ton (metric) (tonne)	1 000.0 *	kilograms

7. Angles

Multiply	by	to obtain
cycle (360°)	6.283	radians
degree	0.017 453	radians
Hertz	6.283	radians/second
rev./minute	0.104 7	radians/second
radians	57.295 8	degrees

8. Force

Multiply	by	to obtain
dynes	10^{-5} *	newton
grams (force)	980.7	dynes
kilogram (force)	9.806 65 *	newtons
newton	10^5 *	dynes
	0.102 0	kilogram (force)
	3.597	ounce (force)
	0.224 8	pound (force)
	7.233 0	poundal
ounce (force)	0.2780	newton
	0.0625 *	pound (force)
pound (force)	16.00 *	ounce (force)
	0.45359	kilogram (force)
	4.448	newtons
ton (force) (short)	2000 *	pounds (force)
	8896	newtons

9. Moment of Torque

Multiply	by	to obtain
ounce (f) inch	0.007061	newton meters
pound (f) inch	1.152	kg cm
	0.1130	newton meters
pound (f) foot	1.356	newton meters
newton meters	0.7376	pound (f) foot

10. Pressure

Multiply	by	to obtain
atmospheres	1.01325 *	bars
	33.90	feet of H_2O
	29.92	inches of Hg
	760.0 *	mm of Hg (torr)
	101.325 *	kN/m^2 (k Pa)
	14.696	pounds/sq. inch
bar	75.01	cm of Hg
	10^5 *	N/m^2 (Pa)
	14.50	pounds/sq. inch
dyne/cm^2	0.1000 *	N/m^2 (Pa)
inches of H_2O	248.84	N/m^2 (Pa)
	0.07355	inches of Hg
kg(f)/cm^2	6.763	pounds/sq. inch
kg(f)/m^2	9.806 65 *	N/m^2
mm of Hg (torr)	133.32	N/m^2
	0.019 33	pounds/sq. inch
	13.595	mm of H_2O
newtons/sq. centimeter	1.450	pounds/sq. inch
N/m^2 (pascal)	1.450 x 10^{-4}	pounds/sq. inch
pounds/sq. foot	0.19242	inches of H_2O
	47.880	N/m^2 (Pa)
pounds/sq. inch	0.068 05	atmospheres
	2.036	inches of Hg
	27.708	inches of H_2O
	68.948	millibars
	703.77	mm of H_2O
	51.72	mm of Hg
	0.689 48	N/cm^2
	6 894.8	N/m^2 (Pa)
	7.031 x 10^{-4}	kg (f)/mm^2

11. Velocity

Multiply	by	to obtain
feet/minute	5.080 *	mm/second
feet/second	0.3048 *	meters/second
inches/second	0.0254 *	meters/second
km/hour	0.6214	miles/hour
knot	0.5144	meters/second
	1.151	miles/hour (U.S.)
meters/second	3.2808	feet/second
	2.237	miles/hour (U.S.)
miles/hour	88.0 *	feet/minute
	0.447 04 *	meters/second
	1.6093	km/hour
	0.8684	knots

12. Acceleration

Multiply	by	to obtain
acceleration of gravity (g)	9.806 65 *	meters/second^2
	32.174	feet/second^2
	386.089	inches/second^2
cm/second^2 (gal)	0.010 *	meters/second^2
feet/second^2	0.3048 *	meters/second^2
inches/second^2	0.025 40 *	meters/second^2

13. Power

Multiply	by	to obtain
ergs/second	10^{-7} *	watts
foot pounds (f)/second	1.356	watts
horsepower (electric)	746.0 *	watts
	76.07	kg m/s
horsepower (U.K.)	745.7	watts
	550	foot pounds/s
BTU/second	1 055.9	watts

14. Energy

Multiply	by	to obtain
BTU (mean)	1055.9	joules
	.2520	kg-calories (mean)
	107.7	kg (f) m
calorie, gram (mean)	4.190	joules
erg	10^{-7} *	joules
foot pound (force)	1.355 8	joules
	.138 25	kg (f) meter
kg (f) m	9.806 65*	joules
watt hour	3 600 *	joules
	3.409	BTU
watt second	1.00 *	joule

15. Temperature

Celsius to kelvin	$K = °C + 273.15$
Celsius to Fahrenheit	$°F = 9/5 °C + 32°$
	$= 1.8 (°C + 40) - 40°$
Fahrenheit to Celsius	$°C = 5/9 (°F - 32°)$
	$= \dfrac{(°F + 40)}{1.8} - 40°$
Fahrenheit to kelvin	$K = 5/9 (°F + 459.67°)$
Fahrenheit to Rankin	$°R = °F + 459.67°$
Rankin to kelvin	$K = 5/9 °R$

16. Electrical

Multiply	by	to obtain
oersted	79.58	ampere/meter
faraday	96 490	coulombs
gauss	10^{-4}*	tesla
gilbert	.7958	ampere turn
maxwell	10^{-8}*	weber

*Exact

B. Mathematics

1. Useful Constants

$\pi = 3.141\ 59$	$\pi^2 = 9.869\ 6$
$2\pi = 6.283$	$(2\pi)^2 = 39.478$
$4\pi = 12.566$	$\dfrac{1}{\pi^2} = 0.101\ 3$
$\dfrac{\pi}{2} = 1.570$	$\sqrt{\pi} = 1.772$
$\dfrac{1}{\pi} = 0.318\ 3$	$\text{Log } \pi = 0.497\ 15$
$\dfrac{1}{2\pi} = 0.159\ 2$	
$\sqrt{2} = 1.414\ 21$	$\sqrt{3} = 1.732\ 0$
$\dfrac{1}{\sqrt{2}} = 0.707\ 1$	$\dfrac{1}{\sqrt{3}} = 0.577\ 3$
$e = 2.718\ 281\ 828$	$\sqrt{10} = 3.162$

2. Trigonometric Relationships

	0°	30°	45°	60°	90°	180°	270°
sine	0	0.500	$\dfrac{\sqrt{2}}{2}$	$\dfrac{\sqrt{3}}{2}$	1.00	0	-1.00
cosine	1.00	$\dfrac{\sqrt{3}}{2}$	$\dfrac{\sqrt{2}}{2}$	0.500	0	-1.00	0
tangent	0	$\dfrac{1}{\sqrt{3}}$	1	$\sqrt{3}$	$+\infty$	0	$-\infty$

3. Quadratic Equation Solution

$$x = \frac{-b \pm \sqrt{b^2 - 4ac}}{2a}$$

where: $ax^2 + bx + c = 0$

4. Decibel Formulae

Power:
$$dB = 10 \log \frac{W}{W_0}$$

Pressure or Voltage:
$$dB = 20 \log \frac{p}{p_0}$$

$$dB = 20 \log \frac{E_1}{E_2}$$

C. Dynamic Measurements

1. Sinusoids (applies only to sinusoids)

rms value	$= 0.707 \times$ peak value
rms value	$= 1.11 \times$ average value
peak value	$= 1.414 \times$ rms value
peak value	$= 1.57 \times$ average value
average value	$= 0.637 \times$ peak value
average value	$= 0.90 \times$ rms value
peak-to-peak	$= 2 \times$ peak value
crest factor	$= \dfrac{\text{peak value}}{\text{rms value}}$ (applies to any varying quantity)

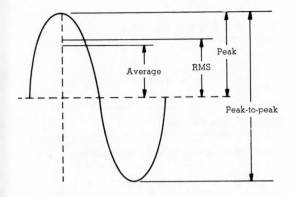

2. Mechanical Impedance

$$Z = \frac{F}{v} = \frac{F}{\omega d} = \frac{\omega F}{a}$$

having units of $\dfrac{\text{lb-sec}}{\text{inch}}$ or $\dfrac{\text{newton-sec}}{\text{meter}} = \text{N/m/s}$

dynamic mass: $Z_a = \dfrac{F}{a}$ dynamic stiffness: $Z_d = \dfrac{F}{d}$

where all terms are phasors, having a magnitude and direction.

3. Displacement, Velocity, Acceleration Relationships

(for sinusoidal motion only)

where: d_0 = peak displacement
D = pk-pk displacement
G = acceleration in g units
f = frequency in Hz
T = period in seconds
g = 9.806 65 m/s^2
= 386.09 in./s^2
= 32.174 ft/s^2

$d = d_0 \sin 2\pi ft$
$v = d_0\, 2\pi f \cos 2\pi ft$
$a = -d_0\, (2\pi f)^2 \sin 2\pi ft$
$G = \dfrac{\text{acceleration}}{g}$

$v_0 = 6.28\, f\, d_0 = 3.14\, f\, D$

$v_0 = 61.4\, \dfrac{G}{f}$ in./s pk

$\quad = 1.560\, \dfrac{G}{f}$ m/s pk

$d_0 = 9.780\, \dfrac{G}{f^2}$ inches pk

$\quad = 0.2484\, \dfrac{G}{f^2}$ meters pk

$G = 0.0511\, f^2 D$
(where: D = inches peak-to-peak)

$G = 2.013\, f^2 D$
(where: D = meters peak-to-peak)

$T = \dfrac{1}{f}$ seconds

4. Motion of a Single-Degree-of-Freedom System

natural frequency: $f_n = \dfrac{\omega}{2\pi} = \dfrac{1}{2\pi}\sqrt{\dfrac{k}{m}} = \dfrac{1}{2\pi}\sqrt{\dfrac{g}{\delta_{st}}}$

where: δ_{st} = static deflection
ω = angular frequency in radians
$k = \dfrac{\text{force}}{\text{deflection}} = \dfrac{m\,a}{d}$

transmissibility: $T = \dfrac{1}{1 - \omega^2/\omega_n^2} = \dfrac{1}{1 - f^2/f_n^2}$
(undamped)

critical damping: $c_c = 2\sqrt{km}$

damping ratio: $\zeta = \dfrac{c}{c_c} = \dfrac{c}{2\sqrt{km}}$

amplification factor: for $\zeta < 0.1$, $Q = \dfrac{1}{2\zeta}$
(at resonance)

5. Resonance Frequency of First Bending Mode

(Unloaded Beams)

$$f_n = C\sqrt{\frac{E\,I\,g}{L^4 W}}$$

where: C = constant, function of method of support
E = elastic modulus
I = moment of inertia of cross section
g = acceleration of gravity
L = length
W = weight per unit length

Support Method	C
Cantilever	0.56
Point support each end	1.57
Both ends fixed	3.56
Totally unsupported	3.56

6. Relative Response of RC Circuit

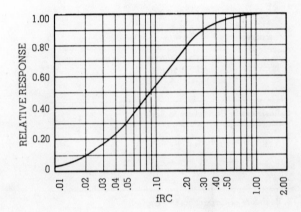

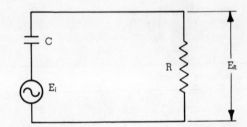

The above curve shows the relative voltage across the load R in a simple RC network as a function of $f \times R \times C$, where:

f = frequency, in hertz
R = load resistance, in ohms
C = capacitance, in farads

D. Electric Circuit Formulae

1. Ohm's Law for DC Circuits:

$$I = \frac{E}{R} = \frac{P}{E} \qquad\qquad R = \frac{E}{I} = \frac{P}{I^2} = \frac{E^2}{P}$$

$$E = IR = \frac{P}{I} \qquad\qquad P = I^2 R = EI = \frac{E^2}{R}$$

where: I = amperes
E = volts
P = watts
R = ohms

2. Ohm's Law for AC Circuits:

$$I = \frac{E}{Z} = \frac{P}{E \cos\theta} \qquad\qquad E = IZ = \frac{P}{I \cos\theta}$$

$$Z = \frac{E}{I} = \frac{P}{I^2 \cos\theta} = \frac{E^2 \cos\theta}{P}$$

$$P = I^2 Z \cos\theta = EI \cos\theta = \frac{E^2 \cos\theta}{Z}$$

where: $\cos\theta = \dfrac{R}{Z} = \dfrac{P}{EI}$ = power factor

θ = angle of lead or lag between current and voltage.
Z = ohms

3. Resistors or Capacitors in Series:

$$R_t = R_1 + R_2 + R_3 + \ldots\ldots\ldots$$

$$\frac{1}{C_t} = \frac{1}{C_1} + \frac{1}{C_2} + \frac{1}{C_3} + \ldots\ldots\ldots$$

4. Resistors or Capacitors in Parallel:

$$\frac{1}{R_t} = \frac{1}{R_1} + \frac{1}{R_2} + \frac{1}{R_3} + \ldots\ldots\ldots$$

$$C_t = C_1 + C_2 + C_3 + \ldots\ldots\ldots$$

ELECTRICAL

5. Two Resistors In Parallel

$$R_t = \frac{R_1 R_2}{R_1 + R_2}$$

$$R_1 = \frac{R_t R_2}{R_2 - R_t}$$

6. Reactance

$$X_C = -\frac{1}{2\pi f C}$$

$$X_L = 2\pi f L$$

7. Impedance

Series Circuit: $Z = \sqrt{R^2 + (X_L + X_C)^2}$

Parallel Circuit: $Z = \dfrac{RX}{\sqrt{R^2 + X^2}}$

8. Quantity of Charge In a Capacitor

$$Q = CE$$

where: Q = charge, in coulombs
E = voltage across capacitor, in volts
C = capacitance, in farads

E. Signal Conditioning

1. Calibration Resistors for Wheatstone Bridges

$$R_c = \frac{R_a}{4}\left(\frac{E_x}{E_o} - 2\right)$$

where: R_c = single shunt calibration resistor, ohms
R_a = bridge arm resistance, ohms
E_x = bridge excitation voltage, volts
E_o = desired change in output voltage, volts

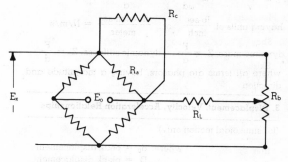

2. Bridge Balance Limiting Resistor

$$R_L \leqq \frac{R_a}{4}\left(\frac{E_x}{E_b} - 2\right)$$

where: R_L = series limiting resistor, ohms
E_b = balance range required, volts

3. Tracking Filter Equations

Time for output of a filter, in response to a step function, to attain 99% of its final value:

$$t = \frac{4}{B} \text{ seconds}$$

where: B = bandwidth of filter at -3dB response
Maximum sweep rate of a narrow band filter for less than 1% error due to filter blurring.

$$R = \frac{B^2}{4} \text{ Hz/second}$$

4. Conversion Charge to Voltage Sensitivity

$$E_s = \frac{1000\, Q_s}{C_p + C_t}$$

where: E_s = voltage sensitivity, mV/g
Q_s = charge sensitivity, pC/g
C_p = transducer capacitance, pF
C_t = total external capacitance, pF

5. Filter Circuits

a. Terminology

1. Bandwidth, for bandpass filters, difference between two frequencies at equal attenuation levels at -3 dB.

2. Cut-off Frequency, f_c, frequency at which response is -3 dB reference midband.

3. Delay Time (Phase Shift), time for a specific point on a signal wave to pass through a filter.

4. Frequency Response, relationship of signal amplitude vs. frequency through filter.

5. Insertion Loss, the essentially constant attenuation in the pass band of the filter.

6. Order (Poles) of a filter, n, the number of reactive components (capacitors or inductors) in the filter.

7. Overshoot, initial increase of output amplitude above final steady state level for a step input.

8. Q-Factor, for band pass filters, $Q = f_o/\Delta f$ where Δf is bandwidth at -3 dB and f_o is center frequency.

9. Rise Time (or Fall Time), time for output to rise (or fall) from 10% to 90% of final magnitude.

10. Shape Factor, for bandpass filters, ratio of bandwidth at -60 dB to bandwidth at -3 dB response.

11. Skirt (Rolloff), terminal slope of attenuation of frequency response, usually expressed as 6n dB/octave, where n is the order of the filter.

12. Transient Response, output of filter for a step input or inpulse input.

b. Filter Types

1. Low-pass, attenuates high frequencies.
2. High-pass, attenuates low frequencies.
3. Band-pass, attenuates both high and low frequencies.
4. Active, includes amplifier.
5. Passive, utilizes only capacitors, inductors, and/or resistors.

6. Driving Long Lines

$$E_o = \frac{I_m}{\sqrt{\dfrac{1}{R_L{}^2} + \dfrac{(2\pi f\,p\,L)^2}{10^{12}}}}$$

E_o = maximum output voltage of signal conditioner, in volts, without exceeding current rating.
I_m = maximum output current, in milliamperes.
R_L = resistive termination at end of cable, in kilo-ohms.
f = maximum frequency, in k Hz.
p = capacitance, in pF per unit length of cable.
L = length of cable in selected units.

F. The Environment

1. Constants

speed of light	C	$= 2.997\ 925 \times 10^{10}$ cm/s
		$= 983.6 \times 10^6$ ft/second
		$= 186\ 284$ miles/second
velocity of sound	v_s	$= 340.3$ meters/second
(dry air at 15° C)		$= 1116$ feet/second
Gravity	g	$= 9.806\ 65$ meters/second
		$= 32.174$ feet/second^2
		$= 386.089$ inches/second^2
1 atmosphere		$= 14.70$ psi
		$= 101.33$ kPa (kN/m^2)
		$= 2116$ lbf /ft^2

2. Acoustic and Vibration Decibels

All quantities are expressed in root-mean-square (rms) values.

dB	Acceleration g	Velocity m/s	Sound Pressure Level in Air Pa (N/m^2)	psi
0	1×10^{-6}	1×10^{-8}	2×10^{-5}	2.90×10^{-9}
20	1×10^{-5}	1×10^{-7}	2×10^{-4}	2.90×10^{-8}
40	1×10^{-4}	1×10^{-6}	2×10^{-3}	2.90×10^{-7}
60	1×10^{-3}	1×10^{-5}	0.02	2.90×10^{-6}
80	.01	1×10^{-4}	0.2	2.90×10^{-5}
100	0.1	1×10^{-3}	2.0	2.90×10^{-4}
120	1.0	0.01	20.	2.90×10^{-3}
140	10.	0.1	200	.0290
160	100.	1.0	2×10^3	.290
180	1000	10	2×10^4	2.90

Reference Levels:
Sound Power
$$P_o = 1\ pW = 10^{-12}\ W = 10^{-5}\ erg/s$$

Airborne Sound Pressure
$$p_{og} = 20\ \mu Pa$$
$$= 0.0002\ \mu bar$$
$$= 0.0002\ dyne/cm^2$$

Waterborne Sound Pressure
$$p_o = 1\ \mu Pa = 10\ \mu bar = 10^{-5}\ dyne/cm^2$$

Acceleration
$$a_o = 1\ \mu g \text{ where}$$
$$g = 9.806\ 65\ m/s^2$$
$$= 386.089\ in/s^2$$

Velocity
$$v_o = 10^{-8}\ m/s = 10^{-6}\ cm/s$$
1 psi rms = 170.8 dB re 20 μPa
1 atmosphere = 14.70 psi

3. Temperature

Fahrenheit to Rankin	$°R = °F + 459.67°$
Celsius to Kelvin	$°K = °C + 273.15°$
Fahrenheit to Celsius	$°C = 5/9\,(°F - 32)$
	$°C = \left(\dfrac{°F + 40}{1.8}\right) - 40$
Celsius to Fahrenheit	$°F = \dfrac{9}{5}\,°C + 32$
	$°F = (°C + 40)\,1.8 - 40$

Appendix 3

Conversion Nomograph for Static Deflection vs Resonance

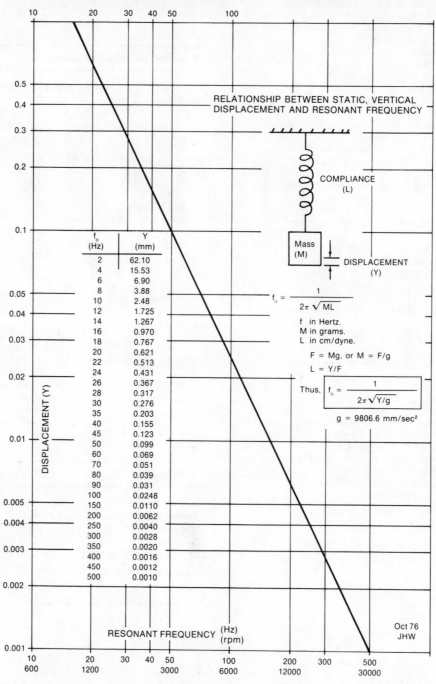

RELATIONSHIP BETWEEN STATIC, VERTICAL DISPLACEMENT AND RESONANT FREQUENCY

COMPLIANCE (L)

Mass (M)

DISPLACEMENT (Y)

f_o (Hz)	Y (mm)
2	62.10
4	15.53
6	6.90
8	3.88
10	2.48
12	1.725
14	1.267
16	0.970
18	0.767
20	0.621
22	0.513
24	0.431
26	0.367
28	0.317
30	0.276
35	0.203
40	0.155
45	0.123
50	0.099
60	0.069
70	0.051
80	0.039
90	0.031
100	0.0248
150	0.0110
200	0.0062
250	0.0040
300	0.0028
350	0.0020
400	0.0016
450	0.0012
500	0.0010

$$f_o = \frac{1}{2\pi\sqrt{ML}}$$

f in Hertz.
M in grams.
L in cm/dyne.

F = Mg, or M = F/g

L = Y/F

Thus, $f_o = \dfrac{1}{2\pi\sqrt{Y/g}}$

g = 9806.6 mm/sec²

DISPLACEMENT (Y)

RESONANT FREQUENCY (Hz) (rpm)

Oct 76
JHW

Courtesy of John H. Woodworth, Bently Nevada Corporation, Frankfurt, Germany.

dB Conversion Chart
for dB Multiplier*

GIVEN: Decibels   TO FIND: Power and Pressure Ratios

TO ACCOUNT FOR THE SIGN OF THE DECIBEL

For positive (+) values of the decibel — Both pressure and power ratios are greater than unity. Use the two right-hand columns.

For negative (—) values of the decibel—Both pressure and power ratios are less than unity. Use the two left-hand columns.

Example—*Given:* ± 9.1 dB. *Find:*

	Power Ratio	Pressure Ratio
+9.1 dB	8.128	2.851
—9.1 dB	0.1230	0.3508

← – dB+ →

Pressure Ratio	Power Ratio	dB	Pressure Ratio	Power Ratio
1.0000	1.0000	0	1.000	1.000
.9886	.9772	.1	1.012	1.023
.9772	.9550	.2	1.023	1.047
.9661	.9333	.3	1.035	1.072
.9550	.9120	.4	1.047	1.096
.9441	.8913	.5	1.059	1.122
.9333	.8710	.6	1.072	1.148
.9226	.8511	.7	1.084	1.175
.9120	.8318	.8	1.096	1.202
.9016	.8128	.9	1.109	1.230
.8913	.7943	1.0	1.122	1.259
.8810	.7762	1.1	1.135	1.288
.8710	.7586	1.2	1.148	1.318
.8610	.7413	1.3	1.161	1.349
.8511	.7244	1.4	1.175	1.380
.8414	.7079	1.5	1.189	1.413
.8318	.6918	1.6	1.202	1.445
.8222	.6761	1.7	1.216	1.479
.8128	.6607	1.8	1.230	1.514
.8035	.6457	1.9	1.245	1.549
.7943	.6310	2.0	1.259	1.585
.7852	.6166	2.1	1.274	1.622
.7762	.6026	2.2	1.288	1.660
.7674	.5888	2.3	1.303	1.698
.7586	.5754	2.4	1.318	1.738
.7499	.5623	2.5	1.334	1.778
.7413	.5495	2.6	1.349	1.820
.7328	.5370	2.7	1.365	1.862
.7244	.5248	2.8	1.380	1.905
.7161	.5129	2.9	1.396	1.950
.7079	.5012	3.0	1.413	1.995
.6998	.4898	3.1	1.429	2.042
.6918	.4786	3.2	1.445	2.089
.6839	.4677	3.3	1.462	2.138
.6761	.4571	3.4	1.479	2.188
.6683	.4467	3.5	1.496	2.239
.6607	.4365	3.6	1.514	2.291
.6531	.4266	3.7	1.531	2.344
.6457	.4169	3.8	1.549	2.399
.6383	.4074	3.9	1.567	2.455
.6310	.3981	4.0	1.585	2.512
.6237	.3890	4.1	1.603	2.570
.6166	.3802	4.2	1.622	2.630
.6095	.3715	4.3	1.641	2.692
.6026	.3631	4.4	1.660	2.754
.5957	.3548	4.5	1.679	2.818
.5888	.3467	4.6	1.698	2.884
.5821	.3388	4.7	1.718	2.951
.5754	.3311	4.8	1.738	3.020
.5689	.3236	4.9	1.758	3.090

← – dB+ →

Pressure Ratio	Power Ratio	dB	Pressure Ratio	Power Ratio
.5623	.3162	5.0	1.778	3.162
.5559	.3090	5.1	1.799	3.236
.5495	.3020	5.2	1.820	3.311
.5433	.2951	5.3	1.841	3.388
.5370	.2884	5.4	1.862	3.467
.5309	.2818	5.5	1.884	3.548
.5248	.2754	5.6	1.905	3.631
.5188	.2692	5.7	1.928	3.715
.5129	.2630	5.8	1.950	3.802
.5070	.2570	5.9	1.972	3.890
.5012	.2512	6.0	1.995	3.981
.4955	.2455	6.1	2.018	4.074
.4898	.2399	6.2	2.042	4.169
.4842	.2344	6.3	2.065	4.266
.4786	.2291	6.4	2.089	4.365
.4732	.2239	6.5	2.113	4.467
.4677	.2188	6.6	2.138	4.571
.4624	.2138	6.7	2.163	4.677
.4571	.2089	6.8	2.188	4.786
.4519	.2042	6.9	2.213	4.898
.4467	.1995	7.0	2.239	5.012
.4416	.1950	7.1	2.265	5.129
.4365	.1905	7.2	2.291	5.248
.4315	.1862	7.3	2.317	5.370
.4266	.1820	7.4	2.344	5.495
.4217	.1778	7.5	2.371	5.623
.4169	.1738	7.6	2.399	5.754
.4121	.1698	7.7	2.427	5.888
.4074	.1660	7.8	2.455	6.026
.4027	.1622	7.9	2.483	6.166
.3981	.1585	8.0	2.512	6.310
.3936	.1549	8.1	2.541	6.457
.3890	.1514	8.2	2.570	6.607
.3846	.1479	8.3	2.600	6.761
.3802	.1445	8.4	2.630	6.918
.3758	.1413	8.5	2.661	7.079
.3715	.1380	8.6	2.692	7.244
.3673	.1349	8.7	2.723	7.413
.3631	.1318	8.8	2.754	7.586
.3589	.1288	8.9	2.786	7.762
.3548	.1259	9.0	2.818	7.943
.3508	.1230	9.1	2.851	8.128
.3467	.1202	9.2	2.884	8.318
.3428	.1175	9.3	2.917	8.511
.3388	.1148	9.4	2.951	8.710
.3350	.1122	9.5	2.985	8.913
.3311	.1096	9.6	3.020	9.120
.3273	.1072	9.7	3.055	9.333
.3236	.1047	9.8	3.090	9.550
.3199	.1023	9.9	3.126	9.772

*From the *Handbook of Noise Measurement,* 7th ed., by Arnold P.G. Peterson and Ervin E. Gross, Jr., 1972, General Radio Company, Concord, Massachusetts.

		−dB+ →					−dB+ →		
Pressure Ratio	Power Ratio	dB	Pressure Ratio	Power Ratio	Pressure Ratio	Power Ratio	dB	Pressure Ratio	Power Ratio
.3162	.1000	10.0	3.162	10.000	.1585	.02512	16.0	6.310	39.81
.3126	.09772	10.1	3.199	10.23	.1567	.02455	16.1	6.383	40.74
.3090	.09550	10.2	3.236	10.47	.1549	.02399	16.2	6.457	41.69
.3055	.09333	10.3	3.273	10.72	.1531	.02344	16.3	6.531	42.66
.3020	.09120	10.4	3.311	10.96	.1514	.02291	16.4	6.607	43.65
.2985	.08913	10.5	3.350	11.22	.1496	.02239	16.5	6.683	44.67
.2951	.08710	10.6	3.388	11.48	.1479	.02188	16.6	6.761	45.71
.2917	.08511	10.7	3.428	11.75	.1462	.02138	16.7	6.839	46.77
.2884	.08318	10.8	3.467	12.02	.1445	.02089	16.8	6.918	47.86
.2851	.08128	10.9	3.508	12.30	.1429	.02042	16.9	6.998	48.98
.2818	.07943	11.0	3.548	12.59	.1413	.01995	17.0	7.079	50.12
.2786	.07762	11.1	3.589	12.88	.1396	.01950	17.1	7.161	51.29
.2754	.07586	11.2	3.631	13.18	.1380	.01905	17.2	7.244	52.48
.2723	.07413	11.3	3.673	13.49	.1365	.01862	17.3	7.328	53.70
.2692	.07244	11.4	3.715	13.80	.1349	.01820	17.4	7.413	54.95
.2661	.07079	11.5	3.758	14.13	.1334	.01778	17.5	7.499	56.23
.2630	.06918	11.6	3.802	14.45	.1318	.01738	17.6	7.586	57.54
.2600	.06761	11.7	3.846	14.79	.1303	.01698	17.7	7.674	58.88
.2570	.06607	11.8	3.890	15.14	.1288	.01660	17.8	7.762	60.26
.2541	.06457	11.9	3.936	15.49	.1274	.01622	17.9	7.852	61.66
.2512	.06310	12.0	3.981	15.85	.1259	.01585	18.0	7.943	63.10
.2483	.06166	12.1	4.027	16.22	.1245	.01549	18.1	8.035	64.57
.2455	.06026	12.2	4.074	16.60	.1230	.01514	18.2	8.128	66.07
.2427	.05888	12.3	4.121	16.98	.1216	.01479	18.3	8.222	67.61
.2399	.05754	12.4	4.169	17.38	.1202	.01445	18.4	8.318	69.18
.2371	.05623	12.5	4.217	17.78	.1189	.01413	18.5	8.414	70.79
.2344	.05495	12.6	4.266	18.20	.1175	.01380	18.6	8.511	72.44
.2317	.05370	12.7	4.315	18.62	.1161	.01349	18.7	8.610	74.13
.2291	.05248	12.8	4.365	19.05	.1148	.01318	18.8	8.710	75.86
.2265	.05129	12.9	4.416	19.50	.1135	.01288	18.9	8.811	77.62
.2239	.05012	13.0	4.467	19.95	.1122	.01259	19.0	8.913	79.43
.2213	.04898	13.1	4.519	20.42	.1109	.01230	19.1	9.016	81.28
.2188	.04786	13.2	4.571	20.89	.1096	.01202	19.2	9.120	83.18
.2163	.04677	13.3	4.624	21.38	.1084	.01175	19.3	9.226	85.11
.2138	.04571	13.4	4.677	21.88	.1072	.01148	19.4	9.333	87.10
.2113	.04467	13.5	4.732	22.39	.1059	.01122	19.5	9.441	89.13
.2089	.04365	13.6	4.786	22.91	.1047	.01096	19.6	9.550	91.20
.2065	.04266	13.7	4.842	23.44	.1035	.01072	19.7	9.661	93.33
.2042	.04169	13.8	4.898	23.99	.1023	.01047	19.8	9.772	95.50
.2018	.04074	13.9	4.955	24.55	.1012	.01023	19.9	9.886	97.72
.1995	.03981	14.0	5.012	25.12	.1000	.01000	20.0	10.000	100.00
.1972	.03890	14.1	5.070	25.70					
.1950	.03802	14.2	5.129	26.30					
.1928	.03715	14.3	5.188	26.92					
.1905	.03631	14.4	5.248	27.54					
.1884	.03548	14.5	5.309	28.18					
.1862	.03467	14.6	5.370	28.84					
.1841	.03388	14.7	5.433	29.51					
.1820	.03311	14.8	5.495	30.20					
.1799	.03236	14.9	5.559	30.90					
.1778	.03162	15.0	5.623	31.62					
.1758	.03090	15.1	5.689	32.36					
.1738	.03020	15.2	5.754	33.11					
.1718	.02951	15.3	5.821	33.88					
.1698	.02884	15.4	5.888	34.67					
.1679	.02818	15.5	5.957	35.48					
.1660	.02754	15.6	6.026	36.31					
.1641	.02692	15.7	6.095	37.15					
.1622	.02630	15.8	6.166	38.02					
.1603	.02570	15.9	6.237	38.90					

		−dB+ →		
Pressure Ratio	Power Ratio	dB	Pressure Ratio	Power Ratio
3.162×10^{-1}	10^{-1}	10	3.162	10
10^{-1}	10^{-2}	20	10	10^2
3.162×10^{-2}	10^{-3}	30	3.162×10	10^3
10^{-2}	10^{-4}	40	10^2	10^4
3.162×10^{-3}	10^{-5}	50	3.162×10^2	10^5
10^{-3}	10^{-6}	60	10^3	10^6
3.162×10^{-4}	10^{-7}	70	3.162×10^3	10^7
10^{-4}	10^{-8}	80	10^4	10^8
3.162×10^{-5}	10^{-9}	90	3.162×10^4	10^9
10^{-5}	10^{-10}	100	10^5	10^{10}

Appendix 5

Phase Shift vs Damping Curve

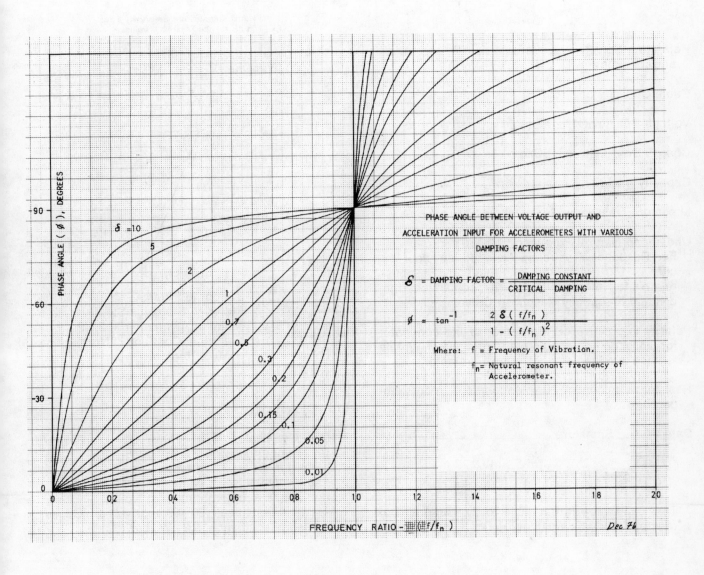

PHASE ANGLE BETWEEN VOLTAGE OUTPUT AND
ACCELERATION INPUT FOR ACCELEROMETERS WITH VARIOUS
DAMPING FACTORS

$$\delta = \text{DAMPING FACTOR} = \frac{\text{DAMPING CONSTANT}}{\text{CRITICAL DAMPING}}$$

$$\phi = \tan^{-1} \frac{2\delta(f/f_n)}{1 - (f/f_n)^2}$$

Where: f = Frequency of Vibration.

f_n = Natural resonant frequency of Accelerometer.

FREQUENCY RATIO - (f/f_n)

Dec 76

Courtesy of John Woodworth, Bently Nevada Corporation, Frankfurt,
Germany.

Appendix 6

Accelerometer Data Sheet and Response

TYPE _____4332_____

Reference Sensitivity _50_ c/s at _23_ °C
and including

Cable Capacity of _105_ pF:

Voltage Sensitivity** _55.8_ mV/g*

Charge Sensitivity _60.1_ pC/g

Capacity (including cable) _1078_ pF

Maximum Transverse Sensitivity at 30 c/s _2.2_ %

Undamped Natural Frequency _48_ kc/s
For Resonant Frequency mounted on steel ex-
citer of 180 grams and for Frequency Response
relative to Reference Sensitivity, see attached individual Frequency Response Curve.

Polarity is positive on the center of the connector
for an acceleration directed from the mounting
surface into the body of the accelerometer.

Resistance minimum 20_{000} Megohms at room temperature.

Date_____ Signature_____

*1 g = 980.6 cm sec^{-2}. $\dfrac{mV}{g} = \dfrac{mV_{rms}}{g_{rms}} = \dfrac{mV_{pk}}{g_{pk}}$

**This calibration is traceable to the National Bu-
reau of Standards Washington D.C.

Individual Temperature Sensitivity Error
In dB rel. the Reference Values.

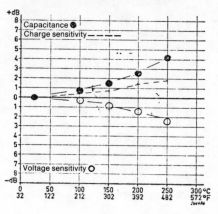

Physical:

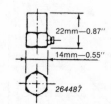

Weight: 30 grams
Material: Stainless Steel
Mounting Thread: 10-32 NF
Electrical Connector: Normal coaxial
10-32 thread.

Environmental:

Humidity: Sealed
Max. Temperature: 260°C or 500°F
Shock Linearity: 3_{000} g typical for 200 μ sec half
sine wave pulse or equivalent
Max. Shock Acceleration: 6_{000} g typical
Magnetic Sensitivity (50-400 cps) < 1 μV/Gauss
Acoustic Sensitiviy < 0.2 μV/μbar

Courtesy of Brueh & Kjaer, Copenhagen, Denmark.

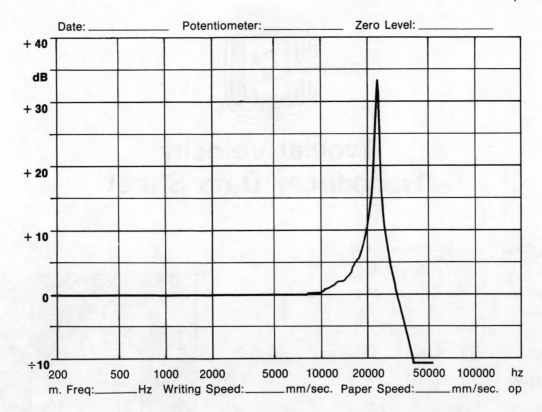

Date: _____   Potentiometer: _____   Zero Level: _____

m. Freq:_____Hz Writing Speed:_____mm/sec. Paper Speed:_____mm/sec. op

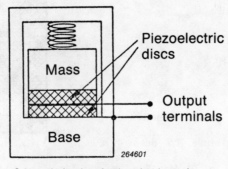

Schematic drawing of a piezoelectric accelerometer.

Appendix 7

Typical Velocity
Transducer Data Sheet

SPECIFICATIONS

Frequency Range: 12 Hz - 1000 Hz ± 8%

-8% ±13% @ 10 Hz

Sensitivity: 764 ± 10% Mv RMS/in/sec (peak)

Temperature Range: –40°F to +500°F

Impedance: R = 2Kohms; L = 0.7 Hy

Damping Factor: 0.7 (critical)

Orientation: Any

Transverse Response: Less than 5% (average) between 10 Hz and 1000 Hz

Magnetic Field Sensitivity: 0.15 inches/sec/gauss

Housing: Waterproof, dustproof

Connector: 2-pin 732S-10SL-4P

Weight: 21 oz

Grounding: Case and signal ground common

Mounting: 1/4 - 28 tapped hole, 1/4 deep

Max Amplitudes: 125 mils pk-pk to 70 Hz; 30g above 70 Hz

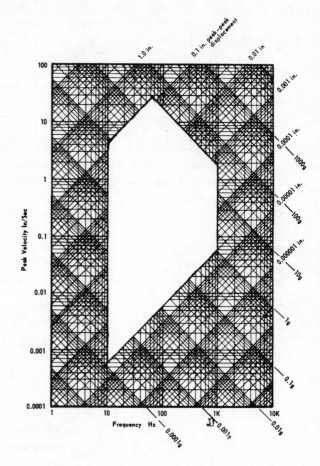

OUTLINE DIMENSIONS

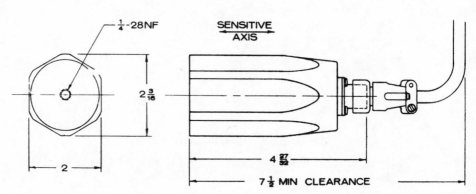

Courtesy of IRD Mechanalysis, Inc., Columbus, Ohio.

Appendix 8

Typical Performance of a Proximity Transducer System

SPECIFICATIONS

PROBE TIP DIAMETERS

5 mm

8 mm (5 mm tip with protective cover)

PROBE LEAD CONSTRUCTION

High Strength Shield — Teflon jacket, 95 ohm impedance, coaxial. Optional armor shielding available, 0.5 or 1.0 meter standard lengths.

PROBE OPERATING ENVIRONMENT

Probe operation unaffected by non-conducting materials such as: oil, gas, steam (below 250°F), plastics, etc., in gap. Functions in most process environments.

EXTENSION CABLE CONSTRUCTION

High-strength Shield-Teflon jacket, 95 ohm impedance, coaxial. Optional armor shielding available. 4.0, 4.5, 8.0, or 8.5 meter standard lengths.

PROXIMITOR OVERALL DIMENSIONS

3.12" x 2.38" x 2.0" (79.2 mm x 60.5 mm x 50.8 mm)

CONNECTORS

Miniature Coaxial — 95 ohm

Optimum Calibration Curves for Various Metals*

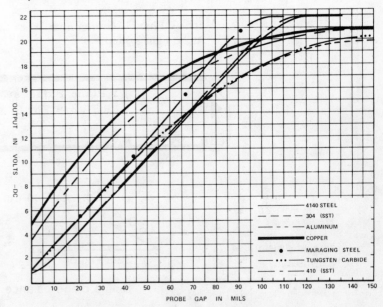

* These curves were determined with a −24 VDC supply, 10,000 ohm load, AISI 4140 Steel Target and at 72°F (22°C) ambient temperature unless otherwise specified.

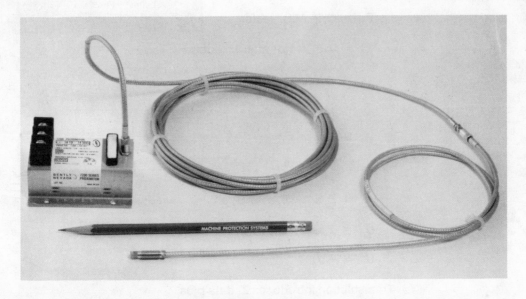

Courtesy of Bently Nevada Corporation, Minden, Nevada.

Appendix 9

Single Plane Balance Program

Follow Phase Convention on Work Sheet. Do not use for seismic sensor-strobe light systems.

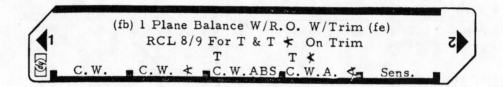

Corr. of Run-out (R.O.) Vector.

Single plane balance by influence coef [a]

Derived from a trial weight (U_T) :

U_B = Unbalance

Z_1 = Orig. V.B. vector

Z_2 = V.B. vector after U_T added.

Z_1 = $a\,U_B$

Z_2 = $a\,(U_B + U_T)$

$Z_2 \cdot Z_1$ = $a\,(U_T)$

a = $Z_2 - Z_1 / U_T$

= $Z_2\,e^{\theta_2} - Z_1\,e^{\theta_1} / U_T\,e^{\theta_T}$

U_B = Z_1/a = $Z_1 e^{\theta_T}/a e^{\theta_T}$

U_a = $-U_B$

Trim solution introduces Z_3 data plus uses C.W. as U_T.

KEY ENTRY	KEY CODE	KEY ENTRY	KEY CODE	KEY ENTRY	KEY CODE	KEY ENTRY	KEY CODE
g LBL b	32 25 12	RCL C	34 13	h RTN	35 22	RCL 7	34 07
R/S	84	RCL 9	34 09	f LBL C	31 25 13	h RTN	35 22
Enter	41	÷	81	RCL C	34 13	f LBL C	31 25 13
R/S	84	ST I	35 33	h RTN	35 22	RCL A	34 11
f → R	31 72	h 1/x	35 62	f LBL D	31 25 14	h RTN	35 22
STO 0	33 00	RCL 5	34 05	RCL D	34 14	f LBL D	31 25 14
h x≷y	35 52	X	71	h RTN	35 22	RCL B	34 12
STO 1	33 01	STO A	33 11	f LBL E	31 25 15	h RTN	35 22
R/S	84	RCL 5	34 05	RCL E	34 15		
Enter	41	÷	81	h RTN	35 22		
R/S	84	STO E	33 15	g LBL e	32 25 15		
f → R	31 72	RC I	35 34	R/S	84		
STO 2	33 02	f P≷S	31 42	Enter	41		
h x≷y	35 52	STO 0	33 00	R/S	84		
STO 3	33 03	f P≷S	31 42	f → R	31 72		
RCL 1	34 01	RCL D	34 14	RCL 0	34 00		
STO(-) 3	33 51 03	RCL 8	34 08	-	51		
RCL 0	34 00	-	51	STO 6	33 06		
STO(-) 2	33 51 02	ST I	35 33	h x≷y	35 52		
RCL 3	34 03	RCL 4	34 04	RCL 1	34 01		
Enter	41	RC I	35 34	-	51		
RCL 2	34 02	-	51	STO 7	33 07		
g → P	32 72	STO B	33 12	RCL A	34 11		
STO 5	33 05	1	01	STO 9	33 09		
h x≷y	35 52	8	08	RCL B	34 12		
STO 4	33 04	0	00	STO 8	33 08		
R/S	84	+	61	f GSB 2	31 22 02		
Enter	41	STO B	33 12	RCL B	34 12		
R/S	84	f (x<0)	31 71	Enter	41		
f → R	31 72	f (GSB) 0	31 22 00	RCL A	34 11		
STO 6	33 06	Enter	41	f → R	31 72		
h x≷y	35 52	3	03	STO 2	33 02		
STO 7	33 07	6	06	h x≷y	35 52		
RCL 0	34 00	0	00	STO 3	33 03		
STO(-) 6	33 51 06	h x≷y	35 52	RCL 8	34 08		
RCL 1	34 01	g (x>y)	32 81	Enter	41		
STO(-) 7	33 51 07	f (GSB) 1	31 22 01	RCL 9	34 09		
R/S	84	STO B	33 12	f → R	31 72		
STO 8	33 08	h RTN	35 22	STO(-) 2	33 51 02		
R/S	84	f LBL 0	31 25 00	h x≷y	35 52		
STO 9	33 09	3	03	STO(-) 3	33 51 03		
f LBL 2	31 25 02	6	06	RCL 3	34 03		
RCL 6	34 06	0	00	Enter	41		
RCL 2	34 02	+	61	RCL 2	34 02		
-	51	h RTN	35 22	g → P	32 72		
STO C	33 13	f LBL 1	31 25 01	STO 6	33 06		
RCL 7	34 07	3	03	h x≷y	35 52		
RCL 3	34 03	6	06	STO 7	33 07		
-	51	0	00	f (x<0)	31 71		
STO D	33 14	-	51	f (GSB) 0	31 22 00		
Enter	41	h RTN	35 22	STO 7	33 07		
RCL C	34 13	f LBL A	31 25 11	h RTN	35 22		
g → P	32 72	RCL A	34 11	f LBL A	31 25 11		
STO C	33 13	h RTN	35 22	RCL 6	34 06		
h x≷y	35 52	f LBL B	31 25 12	h RTN	35 22		
STO D	33 14	RCL B	34 12	f LBL B	31 25 12		

Appendix 10

HP 67 Single Plane Balance Program for Probes, Key Phasor, and Run-out

C. JACKSON

MACHINE IDENTIFICATION _5131-1_ DATE _8/1/78_
RUN-OUT RECORDED @ _300_ R.P.M.; "AT-SPEED" DATA @ _1200_ RPM
KEY PHASOR PROBE IS @ _0_°; PROBES @ _90_° CIRCLE ROTATION

- HP 67 IN "RUN"... PRESS [f][PROGRAM]...READ MAG. CARD SIDES 1 & 2
- PRESS [f][b]

- ENTER *
RUN OUT
DATA, Z_{RO}

	ANGLE		AMOUNT
1	50 R/S °	2	0.7 R/S

- ENTER "AT-SPEED" DATA

- ENTER ORIG. DATA, Z_1

3	44 R/S °	4	3.00 R/S

PLACE TRIAL WEIGHT ON ROTOR.
USE PHASE LOGIC IN PLACING TRIAL WT.

- ENTER T. W. DATA Z_2

5	56 R/S °	6	5.0 R/S

- ENTER T. W. LOCATIONS

7	74 R/S °	8	60 GRAMS UNITS

- PRESS [A,B] TO READ OUT WT. ADD LOCATIONS; [C,D,E] OTHER DATA

	A	B		C	D	E
CORR. WEIGHT + LOCATION	64.10	223.38 °		2.16	72.8 °	27.81 gms/mil
	AMT.	LOC.		T VECTOR	T ANGLE	ROTOR SENSITIVITY, eg. grams/mil.

$T = Z_2 - Z_1$

- PRESS [f][e] ENTER TRIM DATA

- ENTER TRIM DATA (OR FINAL RESULTS)

	ANGLE		AMT
9	46	10	0.8
	TRIM FINAL		TRIM FINAL

- PRESS [A,B] TO READ OUT TRIM CORRECTIONS & [C,D,E] OTHER DATA. [PRESS RCL 8&9 FOR TA & T¢] AFTER TRIM IF NEEDED.

	A	B		C	D	E
TRIM WTS. + LOC.	3.29	200.46 °		67.14	222.29 °	29.13
	AMT.	LOC.		AMT.	LOC.	ROTOR SENS BASED ON C/D
	FOR SEPARATE TRIM WT. ADD.			FOR MOD. WT. ABOVE		

*NUMBERS IN BLOCK [1-10] ARE IN PROGRAM SEQUENCE WITH THE R/S FOLLOWING EACH INPUT

NOTES
- PHASE ANGLES (ALL) TAKEN AGAINST ROTATION FROM VIB. DATA PROBE (SENSOR).
- TRIAL WTS. (T.W.) ARE ALSO PLACED AGAINST ROTATION FROM VIB DATA PROBE.
- CORRECTION WTS. & ANGLES HOLD SAME LOGIC.
- SOLUTION ...NE BASED ON AT ...TIONS

8-10-78

Two Plane Balancing Program, Proximity Probe, CJ Convention, W/R.O. Correction

Use with Worksheet Phase Convention only. This program was adapted from a two plane program with run-out by IRD which was adapted from IRD Application Report 112. It will also do a two vector summing by pressing [f] [c] and entering Vector 1 angle, Vector 1 amount, Vector 2 angle, Vector 2 amount; then pressing "A" for new vector amount and "B" for new vector angle.

1	fa Initialize Runout	fb Initialize 2-Plane	fc Initialize Vect. Add.			2
	A Amt.-N	B ∡-Near	C A-Far	D ∡-Far	E Trim	

KEY ENTRY	KEY CODE
g LBL b	32 25 12
R/S	84
STO 2	33 02
R/S	84
STO 3	33 03
R/S	84
R/S	84
f GSB O	31 22 00
STO 6	33 06
h x≷y	35 52
STO 7	33 07
RCL 2	34 02
RCL 3	34 03
R/S	84
R/S	84
f GSB O	31 22 00
STO 8	33 08
h x≷y	35 52
STO 9	33 09
R/S	84
STO 4	33 04
R/S	84
STO 5	33 05
R/S	84
R/S	84
f GSB O	31 22 00
STO A	33 11
h x≷y	35 52
STO B	33 12
RCL 4	34 04
RCL 5	34 05
R/S	84
R/S	84

KEY ENTRY	KEY CODE
f GSB O	31 22 00
STO C	33 13
h x≷y	35 52
STO D	33 14
RCL 7	34 07
+	61
STO O	33 00
RCL C	34 13
RCL 6	34 06
X	71
STO 1	33 01
RCL 8	34 08
RCL A	34 11
X	71
STO E	33 15
RCL B	34 12
RCL 9	34 09
+	61
RCL E	34 15
RCL O	34 00
RCL 1	34 01
f GSB O	31 22 00
STO O	33 00
h x≷y	35 52
STO 1	33 01
R/S	84
RCL 1	34 01
-	51
STO + 9	33 61 09
RCL D	34 14
+	61
STO D	33 14
R/S	84

KEY ENTRY	KEY CODE
RCL O	34 00
	81
STO X 8	33 71 08
RCL C	34 13
X	71
STO C	33 13
R/S	84
RCL 1	34 01
-	51
STO + 7	33 61 07
RCL B	34 12
+	61
STO B	33 12
R/S	84
RCL O	34 00
÷	81
STO X 6	33 71 06
RCL A	34 11
X	71
STO A	33 11
f LBL 1	31 25 01
RCL 2	34 02
RCL 3	34 03
f P≷S	31 42
RCL 2	34 02
RCL 3	34 03
f GSB O	31 22 00
f P≷S	31 42
CHS	42
STO 3	33 03
h x≷y	35 52
STO 2	33 02
RCL D	34 14

KEY ENTRY	KEY CODE
+	61
STO O	33 00
RCL 3	34 03
RCL C	34 13
X	71
STO 1	33 01
RCL B	34 12
STO + 2	33 61 02
RCL A	34 11
STO X 3	33 71 03
RCL 4	34 04
RCL 5	34 05
f P≷S	31 42
RCL 4	34 04
RCL 5	34 05
f GSB O	31 22 00
f P≷S	31 42
CHS	42
STO 5	33 05
h x≷y	35 52
STO 4	33 04
RCL 7	34 07
+	61
STO E	33 15
RCL 5	34 05
RCL 6	34 06
X	71
h ST I	35 33
RCL 8	34 08
STO X 5	33 71 05
RCL 9	34 09
STO + 4	33 61 04
RCL O	34 00

KEY ENTRY	KEY CODE
RCL 1	34 01
RCL 4	34 04
RCL 5	34 05
f GSB O	31 22 00
STO 4	33 04
h x≷y	35 52
STO 5	33 05
RCL E	34 15
h RC I	35 34
RCL 2	34 02
RCL 3	34 03
f GSB O	31 22 00
STO 2	33 02
h x≷y	35 52
STO 3	33 03
RCL 4	34 04
R/S	84
f LBL O	31 25 00
f R ←	31 72
STO O	33 00
h x≷y	35 52
STO 1	33 01
h R ↑	35 54

KEY ENTRY	KEY CODE
h R ↑	35 54
f R ←	31 72
STO - O	33 51 00
h x≷y	35 52
RCL 1	34 01
h x≷y	35 52
-	51
RCL O	34 00
g →P	32 72
h RTN	35 22
f LBL E	31 25 15
1	01
R/S	84
STO 2	33 02
2	02
R/S	84
STO 3	33 03
3	03
R/S	84
STO 4	33 04
4	04
R/S	84
STO 5	33 05

KEY ENTRY	KEY CODE
GTO 1	22 01
g LBL a	32 25 11
f P≷S	31 42
1	01
R/S	84
STO 2	33 02
2	02
R/S	84
STO 3	33 03
3	03
R/S	84
STO 4	33 04
4	04
R/S	84
STO 5	33 05
f P≷S	31 42
R/S	84
f LBL A	31 25 11
RLC 4	34 04
h RTN	35 22
f LBL B	31 25 12
CHS	42

KEY ENTRY	KEY CODE
RCL 5	34 05
h RTN	35 22
f LBL C	31 25 13
RCL 2	34 02
h RTN	35 22
f LBL D	31 25 14
CHS	42
RCL 3	34 03
h RTN	35 22
g LBL c	32 25 13
R/S	84
R/S	84
CHS	42
R/S	84
R/S	84
f GSB O	31 22 00
STO 4	33 04
h x≷y	35 52
STO 5	33 05
RCL 4	34 04
R/S	84

Appendix 12

HP 67 Two Plane
Balance Program for Probes,
Key Phasor, and Run-out

ROTATION COUNTER-CLOCKWISE

ROTATION CLOCKWISE

MACHINE IDENTIFICATION __99DM-1__ DATE __7/78__
RUN-OUT RECORDED @ __300__ R.P.M.; "AT-SPEED" DATA @ __3600__ RPM
KEY PHASOR PROBE IS @ __0__ °; PROBES @ __90__ °N & __90__ °F

K.P.
VIB. PROBE φ

- H.P. 67 IN "RUN"   PRESS [f][PROGRAM]...READ MAG. CARD SIDES 1 & 2
- PRESS [f][a]

	NEAR PLANE		FAR PLANE	
	ANGLE	AMOUNT	ANGLE	AMOUNT
• ENTER RUN-OUT DATA	1 272° R/S	2 0.5 R/S	3 123° R/S	4 0.4 R/S

- PRESS [f][b] AND ENTER "AT-SPEED" DATA

	ANGLE	AMOUNT	ANGLE	AMOUNT
ORIG. DATA	5 148° R/S	6 1.8 R/S	11 115° R/S	12 3.6 R/S
T.W.@NEAR PLANE	7 178° R/S	8 1.1 R/S	13 98° R/S	14 2.0 R/S
T.W.@FAR PLANE	9 98° R/S	10 2.1 R/S	15 102° R/S	16 3.7 R/S
T.W. LOCATIONS	17 120°	18 4.9 gms UNITS	19 220°	20 4.9 gms UNITS

- PRESS [A,B,C,D] TO READ OUT WT. ADD LOCATIONS

	A	B	C	D
CORR. WEIGHT + LOCATION	7.49 AMT	84.96° LOC	5.32 AMT	179.73° LOC

X

- PRESS [E] ENTER TRIM DATA

	NEAR PLANE		FAR PLANE	
	ANGLE	AMT.	ANGLE	AMT.
	1 268° R/S	2 0.6 R/S	3 0.85° R/S	4 0.5 R/S

- PRESS [A,B,C,D] TO READ OUT TRIM CORRECTIONS

	A	B	C	D
TRIM WEIGHTS + LOCATION	2.79 AMT	-47.11 312.89° LOC	1.55 AMT	135.54° LOC

X

NOTE: AT PTS. (3) MARKED (X→) IN LEFT MARGIN. SYSTEM PARAMETERS MAY BE RECORDED IN RUN BY [f] [W/DATA] AND MAG. CARD.

NOTES
- PHASE ANGLES (ALL) TAKEN AGAINST ROTATION FROM VIB. DATA PROBE (SENSOR).
- TRIAL WTS. (T.W.) ARE ALSO PLACED AGAINST ROTATION FROM VIB DATA PROBE.
- CORRECTION WTS. & ANGLES HOLD SAME LOGIC.

10-10-77

103

Appendix 13

Four Channel
Vector Subtraction (with Printing)

The HP 67 can use this same program by substituting LST X (31-84) for Print X.

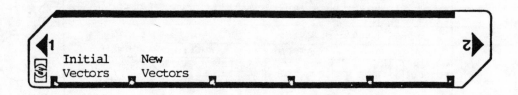

STEP	INSTRUCTIONS	INPUT DATA/UNITS	KEYS		OUTPUT DATA/UNITS
	SAMPLE PROBLEM				
	Label "A" Load Initial Vectors			A	
	01.	I.D.			1.00
A_{o_1} =	1.00 ***	Mils		R/S	1.00
Θ_{o_1} =	60.00 ***	Degrees		R/S	60.00
	02.	I.D.			2.00
A_{o_2} =	1.00 ***	Mils		R/S	1.00
Θ_{o_2} =	180.00 ***	Degrees		R/S	180.00
	03.	I.D.			3.00
A_{o_3} =	1.50 ***	Mils		R/S	1.50
Θ_{o_3} =	275.00 ***	Degrees		R/S	275.00
	04.	I.D.			4.00
A_{o_4} =	2.00 ***	Mils		R/S	2.00
Θ_{o_4} =	355.00 ***	Degrees		R/S	355.00
	Calculator then goes to Label "B"				
	ready to accept the New Vectors				
	11.	I.D.			11.00

Courtesy of Robert C. Eisenmann, consultant, Bently Nevada Corporation, Houston, Texas.

STEP	INSTRUCTIONS	INPUT DATA/UNITS	KEYS		OUTPUT DATA/UNITS
	SAMPLE PROBLEM - Continued				
	A_{11} = 1.90 ***	Mils		R/S	1.90
	Θ_{11} = 75.00 ***	Degrees		R/S	75.00
	12.	I.D.			12.00
	A_{12} = 2.30 ***	Mils		R/S	2.30
	Θ_{12} = 181.00 ***	Degrees		R/S	181.00
	13.	I.D.			13.00
	A_{13} = 1.70 ***	Mils		R/S	1.70
	Θ_{13} = 30.00 ***	Degrees		R/S	30.00
	14.	I.D.			14.00
	A_{14} = 3.00 ***	Mils		R/S	3.00
	Θ_{14} = 355.00 ***	Degrees		R/S	355.00
	Calculator will then print the vectorial difference for each pair of vectors as follows:				
	1.	I.D.			
	A_1 = 0.97 ***	Mils			
	Θ_1 = 90.49 ***	Degrees			
	2.	I.D.			
	A_2 = 1.30 ***	Mils			
	Θ_2 = 181.77 ***	Degrees			
	3.	I.D.			
	A_3 = 2.70 ***	Mils			
	Θ_3 = 60.22 ***	Degrees			
	4.	I.D.			
	A_4 = 1.00 ***	Mils			
	Θ_4 = 355.00 ***	Degrees			

(Appendix 13 continued on next page)

KEY ENTRY	KEY CODE	KEY ENTRY	KEY CODE	KEY ENTRY	KEY CODE	KEY ENTRY	KEY CODE
*LBL A	21 11	DSP 2	-63 02	PRT X	-14	RTN	24
DSP 0	-63 00	R/S	51	X≷Y	-41	*LBL 0	21 00
0	00	PRT X	-14	GSB 0	23 00	X>0?	16-44
1	01	STO 0	35 00	PRT X	-14	RTN	24
PRT X	-14	R/S	51	SPC	16-11	3	03
DSP 2	-63 02	PRT X	-14	RCL 3	36 03	6	06
R/S	51	STO 1	35 01	RCL 2	36 02	0	00
PRT X	-14	SPC	16-11	GSB 1	23 01	+	-55
STO 0	35 00	DSP 0	-63 00	RCL 3	36 03	RTN	24
R/S	51	1	01	RCL 2	36 02	*LBL 1	21 01
PRT X	-14	2	02	GSB 2	23 02	→ R	44
STO 1	35 01	PRT X	-14	DSP 0	-63 00	STO 8	35 08
SPC	16-11	DSP 2	-63 02	2	02	X≷Y	-41
DSP 0	-63 00	R/S	51	PRT X	-14	STO 9	35 09
0	00	PRT X	-14	DSP 2	-63 02	P≷S	16-51
2	02	STO 2	35 02	R↓	-31	RTN	24
PRT X	-14	R/S	51	PRT X	-14	*LBL 2	21 02
DSP 2	-63 02	PRT X	-14	X≷Y	-41	→ R	44
R/S	51	STO 3	35 03	GSB 0	23 00	P≷S	16-51
PRT X	-14	SPC	16-11	PRT X	-14	ST-8	35-45 08
STO 2	35 02	DSP 0	-63 00	SPC	16-11	X≷Y	-41
R/S	51	1	01	RCL 5	36 05	ST-9	35-45 09
PRT X	-14	3	03	RCL 4	36 04	RCL 9	36 09
STO 3	35 03	PRT X	-14	GSB 1	23 01	RCL 8	36 08
SPC	16-11	DSP 2	-63 02	RCL 5	36 05	→ P	34
DSP 0	-63 00	R/S	51	RCL 4	36 04	RTN	24
0	00	PRT X	-14	GSB 2	23 02	R/S	51
3	03	STO 4	35 04	DSP 0	-63 00		
PRT X	-14	R/S	51	3	03		
DSP 2	-63 02	PRT X	-14	PRT X	-14		
R/S	51	STO 5	35 05	DSP 2	-63 02		
PRT X	-14	SPC	16-11	R↓	-31		
STO 4	35 04	DSP 0	-63 00	PRT X	-14		
R/S	51	1	01	X≷Y	-41		
PRT X	-14	4	04	GSB 0	23 00		
STO 5	35 05	PRT X	-14	PRT X	-14		
SPC	16-11	DSP 2	-63 02	SPC	16-11		
DSP 0	-63 00	R/S	51	RCL 7	36 07		
0	00	PRT X	-14	RCL 6	36 06		
4	04	STO 6	35 06	GSB 1	23 01		
PRT X	-14	R/S	51	RCL 7	36 07		
DSP 2	-63 02	PRT X	-14	RCL 6	36 06		
R/S	51	STO 7	35 07	GSB 2	23 02		
PRT X	-14	SPC	16-11	DSP 0	-63 00		
STO 6	35 06	SPC	16-11	4	04		
R/S	51	RCL 1	36 01	PRT X	-14		
PRT X	-14	RCL 0	36 00	DSP 2	-63 02		
STO 7	35 07	GSB 1	23 01	R↓	-31		
SPC	16-11	RCL 1	36 01	PRT X	-14		
SPC	16-11	RCL 0	36 00	X≷Y	-41		
P≷S	16-51	GSB 2	23 02	GSB 0	23 00		
*LBL B	21 12	DSP 0	-63 00	PRT X	-14		
DSP 0	-63 00	1	01	SPC	16-11		
1	01	PRT X	-14	SPC	16-11		
1	01	DSP 2	-63 02	SPC	16-11		
PRT X	-14	R↓	-31	SPC	16-11		

Appendix 14

Balance Weight
Hole Splitting Program

This program allows one to vectorial divide a balance correction weight vector proportionately into two existing balancing holes (locations). It is written for HP 67 language, however, insertion of "spaces" at the indicated arrows and using HP 97 notation, this program is easily converted to the HP 97 format.

BALANCE WEIGHT SPLITTING BETWEEN 2 HOLES

1 Hole   Max Wt
   Splitting Hole

STEP	INSTRUCTIONS	INPUT DATA/UNITS	KEYS	OUTPUT DATA/UNITS
1.	Turn Calculator "ON" and "RUN"			0.00
2.	Load Program - Read Side 1			0.00
3.	Label "A" Hole Splitting - Load Data		A	0.00
	Desired Weight Addition*(	grams	R/S	
	(Wait For Blinking To Stop On Each Entry)			
	Desired Weight Angle	Degrees	R/S	
	First Balance Hole From Probe	Degrees	R/S	
	Second Balance Hole From Probe	Degrees	R/S	
	HP-67 will store each entry on the three angles.			
	Unit then skips 2 spaces and displays in succession			
	Weight @ Hole *** (DSP 2)	grams		A
	Angle Of Hole *** (DSP 0)	Degrees		∝
	Calculator skips 1 space			
	Weight @ Hole *** (DSP 2)	grams		B
	Angle Of Hole *** (DSP 0)	Degrees		β
	Calculator skips 1 space			
	Total Weight Required *** (DSP 2)	grams		A + B
4.	Label "B" to calculate the maximum weight per		B	
	hole, and fraction of weight required for each			
	hole.			
	Load the following data:			
	Hole Diameter *** (DSP 2)	inches	R/S	
	Hole Length *** (DSP 2)	inches	R/S	
	HP-67 displays:			
	Maximum Weight per Hole	grams		Max. Wt.

STEP	INSTRUCTIONS	INPUT DATA/UNITS	KEYS		OUTPUT DATA/UNITS
	SAMPLE PROBLEM				
	Weight A/Maximum Weight *** (DSP 2)	gm/gm			A/Max. Wt
	Weight B/Maximum Weight *** (DSP 2)	gm/gm			B/Max. Wt
	NOTE:				
	Label "A" can use any units for the Weight W,				
	i.e., grams, oz, etc. However, the second				
	part of Label "B" assumes that the calculations				
	made in Label "A" carry the units of grams.				
	Label "A" Hole Splitting - Load Data			A	
	W = 25.00 ***	grams		R/S	25.00
	Θ = 50. ***	Degrees		R/S	
	∝ = 30. ***	Degrees		R/S	
	β = 60. ***	Degrees		R/S	
	Calculator then skips and displays the results				
	as follows: (8 Blinks Each)				
	A = 8.68 ***	grams			8.68
	∝ = 30. ***	Degrees			30.
	B = 17.10 ***	grams			17.10
	β = 60. ***	Degrees			60.
	A + B = 25.78 ***	grams			25.78
	Label "B" Maximum Weight per Hole			B	
	d = 0.50 ***	inches		R/S	
	L = 1.00 ***	inches		R/S	
	Max. Wt. = 25.25 ***	grams			25.25
	A/Max. Wt. = 0.34 ***	gm/gm			0.34
	B/Max. Wt. = 0.68 ***	gm/gm			0.68

KEY ENTRY	KEY CODE	KEY ENTRY	KEY CODE	KEY ENTRY	KEY CODE	KEY ENTRY	KEY CODE
f LBL A	31 25 11	STO 9	33 09	X	71	÷	81
R/S	84	DSP 2	23 02	-	51	h π	35 73
LST X	31 84	RCL 7	34 07	RC I	35 34	X	71
STO 0	33 00	RCL 8	34 08	X	71	RB	84
R/S	84	X	71	STO B	33 12	LST X	31 84
DSP 0	23 00	RCL 6	34 06	RCL A	34 11	X	71
LST X	31 84	RCL 9	34 09	+	61	1	01
STO 1	33 01	X	71	STO C	33 13	2	02
SIN	31 62	-	51	RCL A	34 11	8	08
STO 4	33 04	RCL 0	34 00	LST X	31 84	.	83
RCL 1	34 01	h x y	35 52	RCL 2	34 02	6	06
COS	31 63	÷	81	DSP 0	23 00	X	71
STO 5	33 05	ST I	35 33	LST X	31 84	STO D	33 14
R/S	84	RCL 5	34 05	DSP 2	23 02	LST X	31 84
LST X	31 84	RCL 8	34 08	RCL B	34 17	RCL A	34 11
STO 2	33 02	X	71	LST X	31 84	RCL D	34 14
SIN	31 62	RCL 4	34 04	RCL 3	34 03	÷	81
STO 6	33 06	RCL 9	34 09	DSP 0	23 00	LST X	31 84
RCL 2	34 02	X	71	LST X	31 84	RCL B	34 12
COS	31 63	-	51	DSP 2	23 02	RCL D	34 14
STO 7	33 07	RC I	35 34	RCL C	34 13	÷	81
R/S	84	X	71	LST X	31 84	LST X	31 84
LST X	31 84	STO A	33 11	h RTN	35 22	h RTN	35 22
STO 3	33 03	RCL 4	34 04	f LBL B	31 25 12		
SIN	31 62	RCL 7	34 07	R/S	84		
STO 8	33 08	X	71	LST X	31 84		
RCL 3	34 03	RCL 5	34 05	g X2	32 54		
COS	31 63	RCL 6	34 06	4	04		

Program Description, Equations, Variables, etc.

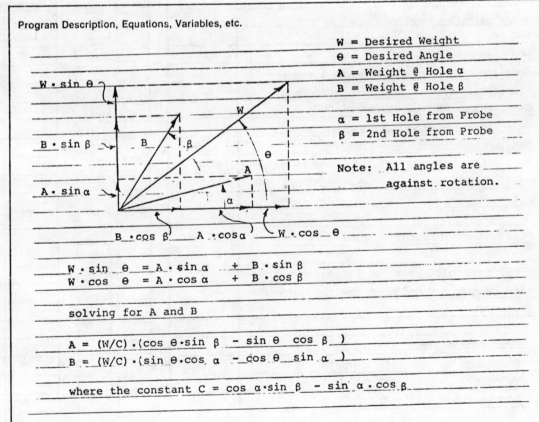

W = Desired Weight

Θ = Desired Angle

A = Weight @ Hole α

B = Weight @ Hole β

α = 1st Hole from Probe

β = 2nd Hole from Probe

Note: All angles are against rotation.

$$W \cdot \sin \theta = A \cdot \sin \alpha + B \cdot \sin \beta$$
$$W \cdot \cos \theta = A \cdot \cos \alpha + B \cdot \cos \beta$$

solving for A and B

$$A = (W/C) \cdot (\cos \theta \cdot \sin \beta - \sin \theta \cos \beta)$$
$$B = (W/C) \cdot (\sin \theta \cdot \cos \alpha - \cos \theta \sin \alpha)$$

where the constant $C = \cos \alpha \cdot \sin \beta - \sin \alpha \cdot \cos \beta$

Operating Limits and Warnings

Appendix 15

Vibration
Conversion Program

This program is written for HP 97, but by inserting "LST X" for "Print X" direct displays on HP 67 are possible. The frequency can be re-entered and new conversions made repeatedly.

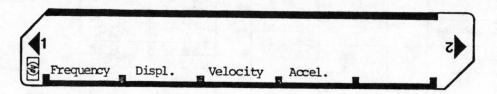

Frequency   Displ.   Velocity   Accel.

STEP	INSTRUCTIONS	INPUT DATA/UNITS	KEYS		OUTPUT DATA/UNITS
1.	Turn Calculator "ON" and "RUN"				0.00
2.	Load Program – 1 side only				0.000
3.	Label "A"			A	0.000
	Load Frequency	cpm		R/S	F
4.	Select the known vibration units i.e.,				
	(1).Displacement known, and velocity/accel.				
	desired information use Label "B"			(B) or	
	(2).Velocity known, and displacement/accel.				
	desired information use Label "C"			(C) or	
	(3).Acceleration known, and displacement/velocity·				
	desired information use Label "D"			(D)	
		Mils or			
	Load the known vibration amplitude	IPS or		R/S	
		G's			
	HP-97 then calculates the 2 missing vibra-				
	tion levels, and produces the following				
	print-out:				
	Frequency (cpm) *** (DSP 0)				F
	Displacement (Mils)*** (DSP 3)				D
	Velocity (IPS) *** (DSP 3)				V
	Acceleration (G's) *** (DSP 3)				A

Courtesy of R.C. Eisenmann, consultant, Bently Nevada Corporation, Houston, Texas.

STEP	INSTRUCTIONS	INPUT DATA/UNITS	KEYS		OUTPUT DATA/UNITS
	EXAMPLE: 5,000 cpm & 3 Mils				
	Label "A"			A	0.000
	Load Frequency = 5,000.	cpm		R/S	1.356
	Select Label "B"			B	1.356
	Load Displacement = 3.	Mils		R/S	- -
	Print-Out				
	5,000. *** (Frequency)	cpm			
	3.000 *** (Displacement)	Mils			
	0.785 *** (Velocity)	IPS			
	1.065 *** (Acceleration)	G's			

KEY ENTRY	KEY CODE	KEY ENTRY	KEY CODE	KEY ENTRY	KEY CODE	KEY ENTRY	KEY CODE
*LBL A	21 11	3	03	X	-35	RCL 3	36 03
DSP 3	-63 03	1	01	STO D	35 14	÷	-24
R/S	51	÷	-24	GTO E	22 15	STO C	35 13
STO A	35 11	STO 2	35 02	*LBL C	21 13	*LBL E	21 15
6	06	RCL 0	36 00	R/S	51	RCL A	36 11
0	00	6	06	STO C	35 13	DSP 0	-63 00
÷	-24	1	01	RCL 2	36 02	PRT X	-14
STO 0	35 00	.	-62	÷	-24	DSP 3	-63 03
1	01	4	04	STO B	35 12	RCL B	36 12
3	03	4	04	RCL C	36 13	PRT X	-14
9	09	÷	-24	RCL 3	36 03	RCL C	36 13
.	-62	STO 3	35 03	X	-35	PRT X	-14
8	08	RTN	24	STO D	35 14	RCL D	36 14
5	05	*LBL B	21 12	GTO E	22 15	PRT X	-14
÷	-24	R/S	51	*LBL D	21 14	SPC	16-11
x2	53	STO B	35 12	R/S	51	SPC	16-11
STO 1	35 01	RCL 2	36 02	STO D	35 14	SPC	16-11
RCL 0	36 00	X	-35	RCL 1	36 01	SPC	16-11
3	03	STO C	35 13	÷	-24	RTN	24
1	01	RCL B	36 12	STO B	35 12	R/S	51
8	08	RCL 1	36 01	RCL D	36 14		
.	-62						

Index